Yield and Reliability in Microwave Circuit and System Design

For a complete listing of the *Artech House Microwave Library*,
turn to the back of this book

Yield and Reliability in Microwave Circuit and System Design

Michael D. Meehan and John Purviance

Artech House
Boston • London

Library of Congress Cataloging-in-Publication Data

Meehan, Michael D.
   Yield and Reliability in Microwave Circuit and System Design
   Includes bibliographical references and index.
   ISBN 0-89006-527-6
    1. Microwave integrated circuits—Design and construction—Statistical methods.
2. Engineering design—Statistical methods. 3. Computer-aided design. I. Purviance, John.
II. Title
TK7876.M3892 1993   92-27018
621.381'32—dc20   CIP

British Library Cataloguing in Publication Data

Meehan, Michael D.
   Yield and Reliability in Microwave Circuit and System Design
   I. Title. II. Purviance, John
   621.83182

   ISBN 0-89006-527-6

© 1993 ARTECH HOUSE, INC.
685 Canton Street
Norwood, MA 02062

International Standard Book Number: 0-89006-527-6
Library of Congress Catalog Card Number: 92-27018

10 9 8 7 6 5 4 3 2 1

"There is a need to optimize the unit performance, not at a single parameter value, but over the entire range of parameter values that will be encountered during manufacture."

This book is dedicated to those students, designers, and engineering managers with the courage to learn and to change for the better.

Contents

Foreword
Nightmares

Nightmares are unpleasant and vividly recalled experiences. One nightmare familiar to electronic circuit and design engineers is the discovery that an expensively mass-produced circuit exhibits a performance bearing little more than a passing resemblance to the performance predicted in the laboratory. When a design results in a disastrously low yield caused by a failure to anticipate the effect of component tolerances the customer's specifications are rarely satisfied. Companies have lost millions of dollars through such oversights. (Perhaps the only consolation to electronic engineers is that structural designers suffer from the same dilemma, but compounded by the fact that the collapse of a football stadium or an oil rig is there for all to see; at least the public is spared the sight of "disasters" within an electronic product.)

Designers—and especially their employers—are understandably keen to avoid such nightmare scenarios. Twenty years ago their only source of help was the designer's experience and expertise (still, I hasten to add, of inestimable value). Since that time, a field known as tolerance design has been extensively researched, suitable algorithms have been invented and widely tested, appropriate device modeling techniques devised, and—thankfully—useful, commercially available CAD software put on the market. The field of tolerance design is now mature.

Nevertheless, until the publication of this book, interested designers had to find the time to undertake the not inconsiderable task of sifting through the technical literature to obtain guidance. What they now have available, behind this page, is a presentation in understandable language of the accumulated expertise of Mike Meehan and John Purviance, who over a number of years have worked out how to apply tolerance design to real circuits.

If you are still wondering whether or not to buy this book, a browse through the pages of Chapter 7 may be useful, for you will find there examples which show, in concrete terms, the advantages to be gained from the techniques described in the preceding chapters. The examples include not only circuits such as GaAs MMIC amplifiers but also systems such as a satellite receiver. The earlier chapters present the concepts and techniques used in these real design examples, providing the practical detail needed for successful application. It is not only the words that are important, but the wealth of practical experience and evaluation underlying them in the advice, warnings, and opinions that appear there. The chapter on device and statistical modeling, for example, is the sort of integrated presentation that, to my knowledge, does not exist anywhere else.

If it is your task to design circuits and systems that are "robust" in the sense of being minimally affected by uncontrollable variations in the manufacturing process, then I commend this book to you. To use an expression from Chapter 6: "May your surprises during manufacture be few and far between and hopefully, nonexistent."

Robert Spence
London, 1992

Preface

Even as late as five years ago, computer-aided design of microwave circuits and systems largely involved only single-point optimization. Although Monte Carlo yield analysis was available to designers, few if any used it in the design process. Today, the need for statistical optimization is apparent, due largely to the need for higher yield monolithic microwave integrated circuit designs. About four years ago statistical optimization software became commonly available to the design community. But even with the availability of advanced tools for statistical design, the science is new enough that most designers and engineering managers are not yet fully aware of its benefits or its practice. It is for this reason we undertook the writing of this book.

Statistical design and analysis and its application to microwave circuit and system design are presently an emerging and changing field. There are competing techniques and philosophies on how to properly accomplish the task. For the past eight years the authors have been involved in a particular approach to statistical design involving the development of Monte Carlo-based solutions. Monte Carlo-based analysis and optimization is currently the most general and flexible of all the proposed approaches to statistical design and analysis, and the only technique capable of fully recognizing and exploiting the complexity of realistic random parameter variations (i.e., non-Gaussian variations). For this reason in this book we promote Monte Carlo-based methods.

The approach we have taken in writing this book advocates the proactive use of Monte Carlo methods. The book's style is more a discussion and presentation of our best understanding of the solutions to the statistical design and analysis problem, rather than an encyclopedic presentation of all the possible solutions that exist. In our enthusiasm for the Monte Carlo methods, we may have neglected some details of the broader history of our subject; nevertheless, we have tried to present all methods of statistical design on a "level playing field." Because this is an emerging area, and we are participants in its development, this book depicts the world as seen through our eyes. We hope that the reader will benefit from this viewpoint and the discussions and philosophies presented here.

This book is intended for the design engineer who is just starting to discover the benefits of statistical design and analysis, the seasoned design engineer who wishes to fully understand statistical design, and the engineering manager who wants to evaluate the position and importance of statistical design in the industry. We feel this book, with its background mathematics and its informal style of presentation will be invaluable to all.

While we tried to incorporate "robust" methods for manuscript preparation, we were nonetheless constrained by a tight working schedule. And as chance would have it, there is likely an error or two and potentially even pockets of miscommunication. If the reader identifies any such areas, we welcome correspondence with the aim of improving the overall presentation.

The area of statistical analysis and design will become a mainstay of microwave circuit and system design in the near future, and it is a privilege to present this material to the reader.

Mike Meehan, Thousand Oaks, California
John Purviance, Moscow, Idaho
November 1992

Acknowledgments

Special thanks and love go to my family—Donna, Caitlin, and Nicholas—for their patience, endurance, and understanding. Now I understand why authors pay tribute to their families. Personal thanks are also extended to the anonymous technical reviewer, for he too has exhibited great perseverance. Thanks also to one of the best educators and researchers in the land—John Purviance.

Mike Meehan

I want to thank my beautiful wife Vicki, my two sons Jess and Jason and my daughter Helen for lovingly allowing me to prepare this book, especially when I was away and not the husband and father I should be. I want to thank my dad and mom for their continuing love. Thanks to my "brother" Mike for his friendship. And thanks to Barbara Martin for helping me to keep my eyes on Jesus.

Additionally I want to thank and praise the Lord Jesus Christ for His sacrifice, and His faithfulness and saving grace in my life.

John Purviance

There are many to thank for helping us with this project. First, thanks to Doug Loescher for taking a chance on us and getting us involved in statistical design. Many colleagues and graduate students have helped develop this material and we owe each our thanks. Special thanks to Robert Spence for his pioneering work and his well written papers and books on the subject. We received invaluable assistance in preparing the manuscript from Joan Tozer, Margaret Jenks, and Keith Hoene. Careful review of the manuscript was given by Mitch Mlinar and the Artech reviewer. Their comments and suggestions have greatly changed and improved this book. We are very grateful to EEsof, Inc. and the University of Idaho for providing us with the encouragement and support needed for this project. And finally, thanks to Artech editor Mark Walsh and to Pam Ahl for their help, encouragement, and most of all patience during the writing and editing. We simply had no idea how long it would take.

Chapter 1
Introduction

"Only those firms providing timely delivery of innovative, high-quality, competitively priced products will survive over the long haul."

"There is minimal benefit from optimizing each step in the manufacturing process (design, manufacture, and testing) individually. Instead we need to look at the design process as a whole."

1.1 THE ECONOMICS OF YIELD AND RELIABILITY DESIGN

As the global marketplace becomes a reality for manufacturing firms, survival will require the timely delivery of high-quality, competitively priced products. Localized markets with little competition are quickly disappearing. Outperformed by their Asian and European counterparts, many American firms are only recently restructuring, turning to techniques their competitors have used for years. The catch-all phrase used to describe the modern process of efficient product development and manufacture is *concurrent engineering* [1]. Simply stated, concurrent engineering emphasizes techniques that minimize the risk of undesirable surprise during the life of a product. To accomplish this, all departments responsible for the product, including vendors and consumers, work from product inception to product retirement. Because everybody knows everyone else's expectations, there are no surprises. Even if something undesirable occurs, the fault is with the entire team and not just one individual or department [2]. Using this team-oriented approach, concurrent-engineered products appear on the market earlier, stay out of the repair shop longer, and build consumer confidence continually.

Of all departmental relationships involved in concurrent engineering (see Figure 1.1), none is more important than the one between the design department and the manufacturing department. No amount of collaboration with the other departments can save a design which has not been properly matched to the fabrication process. This is where yield and reliability enter into the economic picture.

Design for high yield and reliability is closely related to robust design (another term to receive heavy use recently). At the root of robust design is the concept of incorporating information about uncertainty into all design stages. As we shall see, sources of uncertainty are many, but significant uncertainty comes from fabrication imprecision, component value variations, environmental effects, and component aging. Robust design aims to "design in" quality by adjusting product design parameters so that uncontrollable variations in *any* product parameter result in minimal deviation from expected product performance. A relatively uncomplicated extension of current design methods, robust design offers a high potential for payback and return. Unfortunately, the number of firms practicing even superficial robust-design techniques are few. Those who substantially invest in robust design should realize in short order sizable boosts in their ability to produce quality products.

One subtle side effect of robust design remains for discussion. Consider the activities involved in concurrent engineering (see Figure 1.2). While we do not go into these techniques here, they mostly involve the implementation of "common sense" in the design process, usually to the effect of attracting each worker's attention to the concept of quality [1, 3–5]. This is intended to promote a team-oriented environment where each team member strives to improve product quality

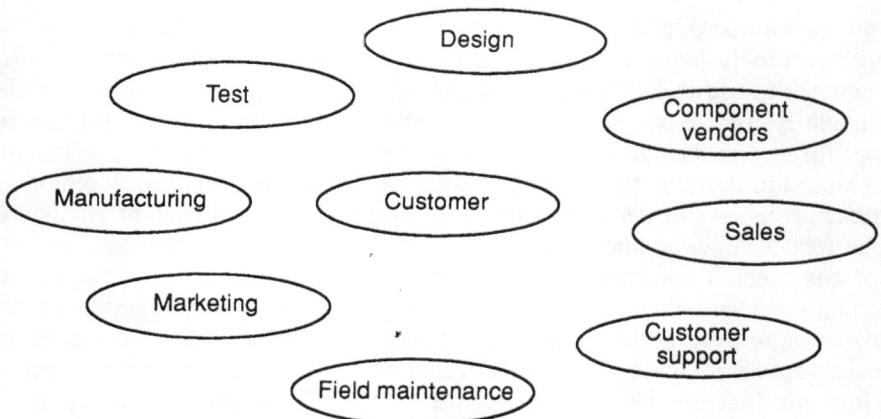

Figure 1.1 Departments playing key roles in concurrent engineering.

by improving his or her own contribution to the overall product. However, winning worker loyalty takes time and must be accomplished with care. Herein lies an additional benefit of robust-design techniques: without undue threat, they serve to unite the two most important departments in the scheme of concurrent engineering—the design department and the manufacturing department. It is no secret that Japanese design engineers rotate jobs with their manufacturing counterparts in an effort to integrate the two departments. They realize the importance of communication and understanding between design and manufacturing. Robust-design techniques, as introduced in this book, offer participating industries the opportunity to create unity, communication, and other mutually beneficial relations between the often separated design and manufacturing departments.

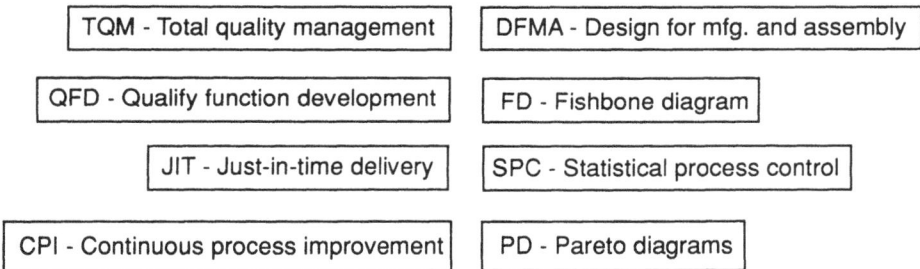

TQM - Total quality management	DFMA - Design for mfg. and assembly
QFD - Qualify function development	FD - Fishbone diagram
JIT - Just-in-time delivery	SPC - Statistical process control
CPI - Continuous process improvement	PD - Pareto diagrams

Figure 1.2 Activities and methods involved in concurrent engineering.

1.2 BACKGROUND TERMINOLOGY AND SCOPE

Robust design and parametric yield and reliability design are terms closely related in meaning. Another term, *tolerance design,* is often used in place of yield and reliability design. We choose not to use this term because it implies limitations that are not actually inherent to the process we describe. In most contexts, the word "tolerance" describes an interval of allowed variation; however, tolerance is only one of the parameter variations typically encountered. A tolerance interval does not indicate *how* the variable changes within this interval or describe the statistical relations that can exist among all of the variables in a design. Hence, the problem of dealing with parameter variations involves more than just tolerances—it is really *statistical* in nature. Rather than using just parameter tolerances to solve this general problem, designers must use the parameter joint probability density function as a statistical model. Although we agree that the term "statistical" is a bit unfriendly, it does more completely describe the viewpoint that designers must adopt to design for high yield.

To help sort out the terminology found in the robust-design literature, refer to Figure 1.3. Robust design, listed at the top level, is concerned with three issues: reliability, product cost, and ease of manufacture. These three concerns are closely related. For example, if you can improve ease of manufacture, the product cost should be reduced. Likewise, if you can make product performance impervious to uncontrollable variations in the product's functional parameters, you are likely to improve both reliability and manufacturability, ultimately decreasing product cost. As we see it, robust design is the process of intelligently trading between the three design concerns of reliability, cost, and manufacturability.

No matter what method you use to accomplish robust design, it will likely include statistical methods or measures (i.e., statistical design). Statistical design is used to minimize the variation in expected product performance in the presence of uncertainties and variations in functional parameters.

In this book we describe general statistical-design techniques that can be applied to many different applications. The two example areas mainly used here are microwave circuits and systems, but, in fact, statistical design can benefit any application that has complex relations between design variables and performance in the presence or design parameter uncertainty. Other application areas for statistical design range from agriculture and automobiles to plastics and petroleum.

To aid in the general application of our ideas, we have developed a nomenclature (illustrated in Figure 1.4) for discussing the examples presented in this book. The final product of any design effort is generically labeled a *unit*. In circuit design, the unit produced is referred to as a circuit; in system design, it is a system. (In mechanical design, a bridge or a gear box might be used as an example unit.) The unit is made up of *components*. For example, components of circuits are resistors, inductors and capacitors; amplifiers and mixers are components of systems. Finally, components are described in terms of *parameters*. For instance, in

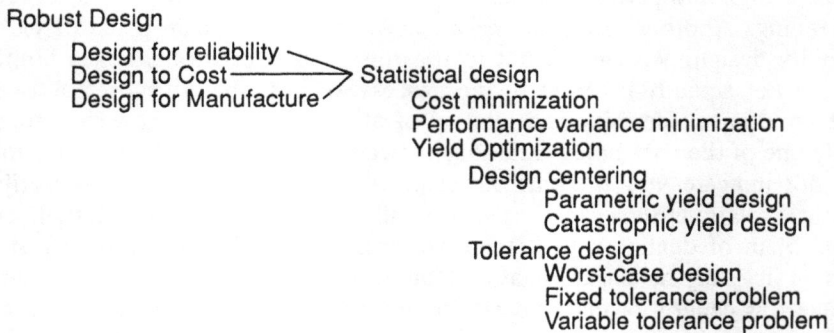

Robust Design
    Design for reliability
    Design to Cost ————> Statistical design
    Design for Manufacture   Cost minimization
                                Performance variance minimization
                                Yield Optimization
                                    Design centering
                                        Parametric yield design
                                        Catastrophic yield design
                                    Tolerance design
                                        Worst-case design
                                        Fixed tolerance problem
                                        Variable tolerance problem

Figure 1.3 Terminology typically found in the robust-design literature.

circuit design, a resistor may have a parameter value of 1 kΩ ± 10%, or in system design a mixer may have a conversion-gain parameter of 1 dB ± 0.3 dB. For units which undergo a fabrication process, such as an integrated circuit, parameters may include process parameters like temperature, doping density, and diffusion rates, as well as material parameters like mobility and conductivity.

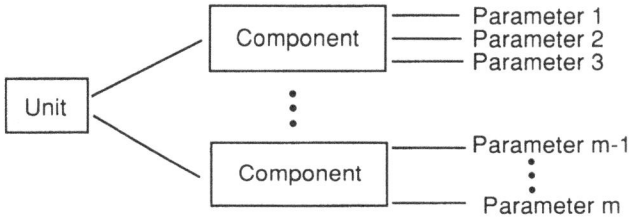

Figure 1.4 Relation of the terms unit, component, and parameter.

1.3 THE DESIGN AND DEVELOPMENT PROCESS

To communicate the motivation behind this book, we will look first at the typical design process presently in use for the general unit, as shown in Figure 1.5. Usually, the intended performance of the unit is given to the designer in terms of a set of performance specifications which must be met after manufacture. In the initial step, the designer visualizes the unit in order to choose a unit structure. From this structure, the designer develops an initial design that he or she then analyzes and optimizes for some performance criteria. The analysis completes the initial design process.

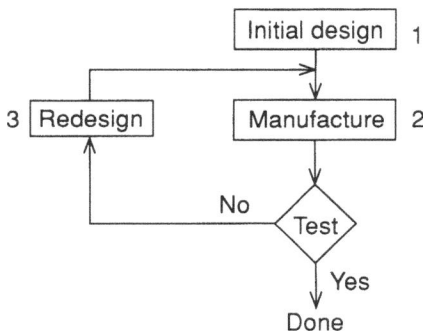

Figure 1.5 The development and design process.

When the designer has confidence that the unit meets performance specifications, the second step begins—the design is released to manufacturing. Units are tested after or during manufacture to make sure they meet specifications. Any unit which fails to meet specifications is rejected. The test specifications are usually the original ones given to the designer at the beginning of the design process.

In practice and with all too frequent regularity, a large percentage of the manufactured units somehow do not meet the test specifications. This is typically due to manufacturing uncertainties, modeling error, or other influences not considered by the designer. When this happens, the third step is initiated—the design is returned to the designer for improvement. Eventually, this iterative process results in a design which, when produced, meets or exceeds all specifications; however, it is often costly and time-consuming. One observer recently used the term "over the wall" to describe this approach to manufacturing [1]. Figure 1.6 shows a cartoon illustrating a traditional over-the-wall design environment with the usual separation between design and manufacturing.

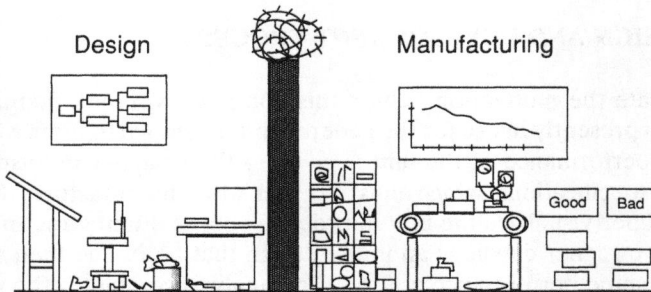

Figure 1.6 Over-the-wall design.

1.4 DESIGN AND DEVELOPMENT FOR MANUFACTURABILITY

A more positive and successful approach to design would be to "get it right the first time," producing reliable, profitable and manufacturable units without the costly iterations. This concept, which includes short cycle time for design-to-production, is called *design for manufacturability*. Some of the first questions posed concerning this approach are:

- How do we design a manufacturable unit?
- How do we eliminate or at least reduce the costly iteration cycle between design and manufacturing?
- How do we design in reliability?

The answers lie in outfitting the design engineer with not only a set of postmanufacture specifications, but also models of the test and manufacturing environment, and specialized tools to make use of those models in conjunction with the unit specifications. Incorporating an accurate statistical model of the manufacturing environment into the design process is the essence of statistical design. In an integrated design environment, where statistical-design methods have been deployed, there is no wall between design and manufacturing. The designer knows the manufacturing environment, and includes it in the initial design process. Figure 1.7 illustrates the improvement.

One of the objectives of this book is to break down the wall between design and manufacture—first, by demonstrating the need for integrated design, and second, by developing methods for communicating between, and thus integrating, the manufacturing environment and the design environment. In the context of Figure 1.7, it is easy to see that the goal of a successful design is to pass the performance test after manufacture, not just to meet the specifications on the designer's workstation. To accomplish this task, the designer needs more information than is generally given in the performance specifications. Accurate models of manufacturing and test processes must be formulated and made available.

In practice there are many different manufacturing and testing environments. Therefore, it would be impractical to model all of them with one general model. In this book we have formulated general testing and manufacturing models that fit most practical situations. In the next two sections we present the models for testing and manufacturing that we will use throughout this book.

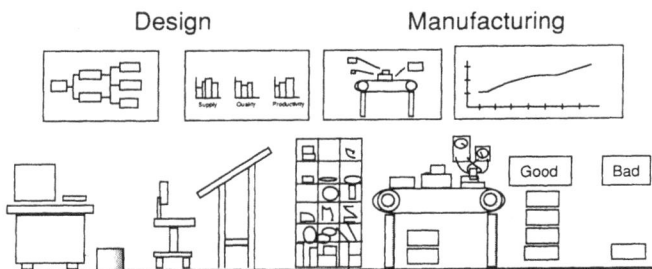

Figure 1.7 Design for manufacture: integrating the manufacturing environment into the initial design.

1.5 TESTING MODEL

Manufactured units are tested to determine if they meet the design specifications originally defined by the unit designer. In our test model, we are concerned with three main issues:

- How are specifications defined and interpreted?
- What failure mechanisms are included?
- What are the sources of testing error, and how are they accommodated?

If the testing environment is to become part of the design process, these kinds of questions must be addressed jointly by design, manufacturing, and test engineers.

1.5.1 Specifications

The unit specifications are a complete description of the performance required of the unit after manufacture. Some typical specifications for a 0.9-MHz < f < 1.1-MHz circuit might be: 9.9 dB < gain < 10.1 dB, noise figure < 2 dB, and 5 dBm < power out < 6 dBm. Specifications are generally given in terms of a window of acceptable values. If the specification for a circuit's gain was simply "10 dB," the circuit would be impossible to manufacture, because this implies the need for 10.000000000... dB of gain. On the other hand, "gain = 10.0 dB ± 0.1 dB" is a realistic specification.

Specifications are always given in terms of a range of acceptable performances, but without the use of statistical design, there is usually no systematic way to incorporate the range into the design process. Statistical design allows the designer to incorporate all of the specification information, including specification ranges.

To see this from another perspective, consider Figure 1.8 where the response of a lowpass filter is plotted versus frequency [6]. The performance specifications are represented by two blocked regions: one for the passband, and one for the stopband. The graph of performance response in (a) is that of the nominal circuit after conventional performance optimization. However, the response in (a) tells us nothing about the response of actual manufactured units during the test. To gain insight into the manufactured response, the design is subject to parameter

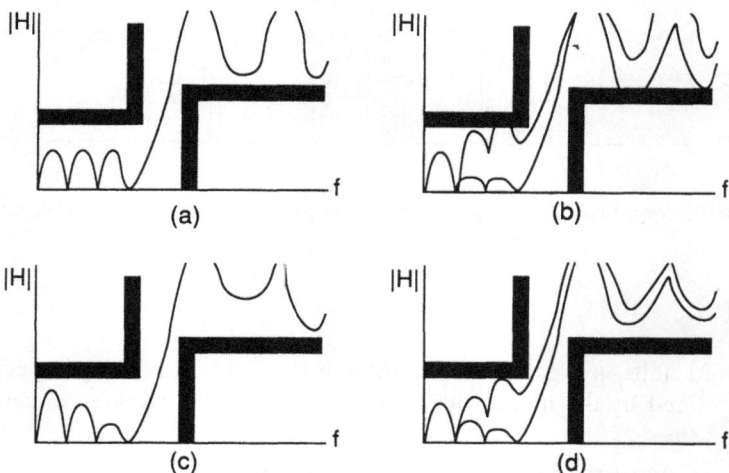

Figure 1.8 The nominal-response and statistical-response envelopes of filter insertion loss (a), (b) before, and (c), (d) after statistical design.

variation representative of the kind found in manufacturing. In (b) we see the statistical-response envelope, and realize that some portion of our lowpass filters are violating the specification. The statistical response as depicted in (b) provides information on the expected performance variation at the end of the manufacturing line. Therefore, the unit statistical response (b) is really more important to the designer than the nominal response (a), and should be the designer's principle concern. In (c), the nominal response is shown after using statistical design. Note how the response has "pulled away" from the specification limits in those ranges where the performance variation is the greatest. The corresponding statistical response (d) shows that all filters meet the performance specification. Like Robin Hood, statistical design robs nominal performance in those regions having little performance variance, to give to regions that are poor statistical performers.

1.5.2 Elements of the Test Model

The testing model we will use is presented in Figure 1.9. It tests the manufactured unit to determine if all specified performance measurements are within specification. Failing units are rejected.

For this model we have made several assumptions. First, it is assumed that there is no sorting or grading of the manufactured units—the units either pass or fail. Second, it is assumed that the units are not tuned during the test. (Even if tuning is allowed, the fixed-test model is useful because it allows the designer to optimally set pretuned performance. This could minimize or even eliminate the need for tuning.)

It is also assumed that there is no measurement noise; that is, the unit performance is measured exactly. Measurement noise can be a problem in certain situations where test-equipment performance is not much better than the performance of the unit under test. Noise adds uncertainty to the measurement process. However, the model presented here appears to fit most modern testing processes.

A more complicated testing model would allow the designer to study how well the units have to be tested; that is, investigate how much money should be spent on the test equipment to ensure an accurate test. The accuracy, expense, and response time of the test equipment could then be factored into the initial unit-design process. This is an important procedure that is not addressed here.

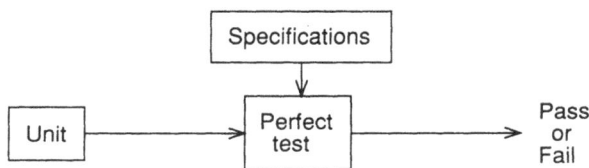

Figure 1.9 Testing model.

A final question regarding the test model deals with sampling: how many tests should be made, and how many units should be tested? To help answer this question, consider our reason for formulating the model in the first place—to implement it on a computer. Once again, we face a trade-off between simulation-model complexity and accuracy. Since the computer affords us the luxury of a large sampling size with minimal errors, our testing model assumes that all units are subject to testing. (Unit sampling also could easily be accommodated within this model.)

The sampling issue will appear again later, in respect to sampling unit parameter statistics and also in the area of simulation-model verification. Let it suffice to remark that in the reality of manufacturing, the upper limit of the number of samples is set by cost considerations, and the lower limit is set by sample-accuracy considerations. Often the latter are sacrificed for the former.

1.5.3 Failure Mechanisms

There are two broad types of failure mechanisms for any manufactured unit:

- Catastrophic failure;
- Parametric failure.

Catastrophic failure accounts for all unit failure due to physical and structural problems within a unit, such as poorly soldered joints, broken components, and material defects. Parametric failure accounts for unit failure due to functional parameter variations during manufacture or unit service life. Although, conceptually unit parametric and catastrophic failures can both be accounted for in statistical design, their origins and remedies are usually not related. Therefore, the separation of the two problems is usually justified and we will only consider parametric failures. For example, if a filter is within specification when $R > 1000\Omega$ and out of specification when $R < 1000\Omega$, in a unit manufactured with $R = 999\Omega$, the failure would be parametric. We assume that all parts used in the manufacture of the units are functional, that all manufacturing processes are mechanically accurate, and that the only uncertainty is the parameter values used during manufacture. Catastrophic failures are not modeled.

1.6 MANUFACTURING MODEL

Our manufacturing model is essentially parametric, and deals with the assembly of the unit using components. In this model the particular unit's performance is uncertain because the parameter values are uncertain. Managing this performance uncertainty in terms of the parameter value uncertainties is the goal of statistical design. A model of the manufacturing process used in this book appears in Figure 1.10.

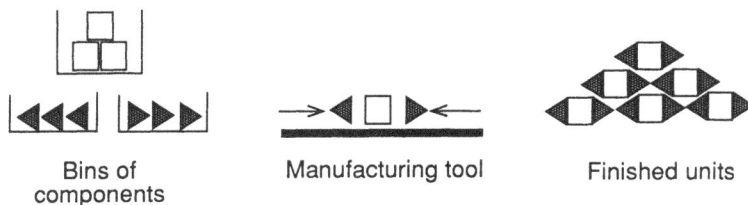

Figure 1.10 Manufacturing-process model.

In Figure 1.10, components are chosen from conceptual storage bins of available parts. The bins contain the components available during manufacture. For instance, if the unit is manufactured from 25 different components, then the manufacturing model would contain 25 bins. The unit is then manufactured from components chosen at random from each of the 25 bins. If a 1-kΩ resistor is needed in the design, there will be a bin filled with 1-kΩ resistors. However, there may not be any true 1-kΩ resistors in the bin. If we were to measure each resistor in the bin we might, in fact, find a 0.995-kΩ resistor and a 1.026-kΩ one, but it is statistically unlikely that the bin will contain a 1.0000000-kΩ resistor. Nonetheless we will characterize the bin with a label: the resistor's *nominal* value will be 1 kΩ.

This example introduces how the manufacturing environment is random, and therefore uncertain, to the designer. Both lumped and distributed component parameters can be subject to uncertainty, and both fit into our conceptual-bin model. For distributed parameters like transmission line width, we have a manufacturing bin containing the transmission-line-width parameter. The line could be characterized by choosing a width parameter from one bin and a length parameter from another bin. (If the length and width are correlated, then the second parameter value is dependent on the value of the first. Our bin model does not accommodate this situation very well. However, we will often use the bin model for intuitive guidance, realizing that it has its limitations.)

To test our idea, we bought three hundred 1.8-kΩ ± 5% resistors and measured their values. Figure 1.11 shows a histogram of the actual tested values for the purchased resistors. As you can see, this demonstrates the unavoidable parametric environment in which any manufactured unit must perform. Modeling the parametric environment is an often overlooked step in today's design process. One focus of this book is to demonstrate to the reader the importance of correct parameter models that accurately describe the statistical component parameter variation encountered in manufacturing.

Recall that our manufacturing model assembles units from components. For example, if the unit is a filter circuit, the components might be resistors, capacitors, and inductors. The important characteristics of each component are its parameters.

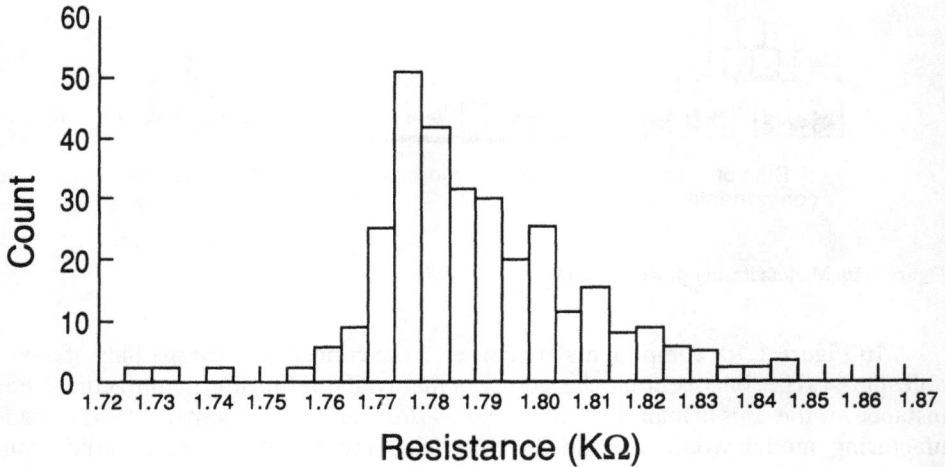

Figure 1.11 Histogram showing measured values obtained from three-hundred 1.8-kΩ ± 5% resistors.

If the component is a low-noise amplifier, the parameters might be S-parameters, the noise figure, and the associated noise resistance. If the component is a microwave FET used in a monolithic microwave integrated circuit (MMIC), the parameters could be circuit-model parameters like C_{gs} and g_m or process-model parameters like gate width, doping density, and mobility, depending on the level of device model being used. No matter what the application, each component is usually described nominally—a "1-kΩ" resistor or "30°K" low-noise amplifier.

In the typical design process, the design engineer selects the unit structure, the components, the process environment, and the parameter nominal values. But as we have seen, this over-the-wall design process neglects to include the manufacturing parametric dynamics and is an invitation for unwanted surprise. It makes little sense to model each component using only its nominal value. In fact, given the evidence of the 1.8-kΩ-resistor value histogram in Figure 1.11, we see that in practice the nominal value is actually not the most likely to occur. (For discrete graded components where the higher grade parts are removed from the total manufactured population, the lower grade components can have a *bimodal* or two-hump distribution with the nominal value even less likely to occur.)

An excellent way to mathematically track the parameter variation is to model each parameter as a random variable. Hence, each parameter of our manufacturing model is described with a probability density function. All the parameters for a unit can thus be modeled with a single *joint* probability density function. This joint probability density function is a complete description of the "randomness" of all of the parameters. (Chapter 6 discusses in detail a process for the successful modeling of component parameter statistics.)

1.6.1 Definition of High Yield

The end result of any manufacturing process is a collection of completed units. One significant goal of the design and manufacturing process is to have the largest percentage of these manufactured units pass performance testing. A measure we use to evaluate the success of the integrated design and manufacturing process is called unit yield. In the following chapter, yield and its calculation will be discussed in detail. However, at this juncture we can define yield as the ratio of the number of units which pass test to the total number of units tested, as the number of units tested becomes very large. When convenient, yield is expressed as a percentage. If all manufactured units pass the performance specification, then the yield is 100%. A 50% yield means that only one half of the units passed performance specification testing. High yield is always a good design goal.

1.6.2 Ways To Achieve High Yield

There are at least three ways to increase the yield of any design:

- Tightly control the manufacturing parameter variations;
- Use a design structure whose performance is insensitive to the uncontrollable manufacturing parameters;
- Choose the nominal parameter values for the given design structure best suited to the particular manufacturing environment.

In this book, we will assume that the manufacturing parameter variations are under "reasonable" control. The results of a careful statistical unit design and sensitivity study may be that certain manufacturing parameters do not need to be controlled tightly; for example, buying 10% resistors instead of 1% resistors may be acceptable.

Design structure can have a significant effect on unit manufacturability. Very little work has been done in this area. One exception is in microwave matching circuits where a systematic method for choosing matching-circuit structure for simple lumped-element matching circuits was presented [7]. Still, additional work needs to be done in this area, and in this book we assume that proper structure selection has been accomplished. (Currently this is typically accomplished by means of trial-and-error statistical analysis on each of the competing structures.)

Our work will concentrate on the proper choice of the nominal parameter value given a particular manufacturing environment. The examples presented at the end of this chapter show that the choice of nominal parameter values can have dramatic effects on the manufacturability (i.e., yield) of a unit.

1.6.3 Parameter Aging and Environmental Model

Finally, it is useful to examine the generality of the model presented here. The manufacturing model shown in Figure 1.10 can be modified conceptually to describe the effects on parameters due to aging or environment. The result of the manufacturing process is many completed units. Each unit is constructed with different parameter values. The goal of good design and manufacture is to have high yield—good performance from many of the completed units. If we are concerned with how a unit will perform over time, we require a single unit to perform well as the parameters vary with time. If we consider the collection of manufactured units in Figure 1.10 to represent a single unit at different points in its lifetime, we can study the effects of aging. For instance, if it is known that a bipolar transistor's beta degrades with time, we can include this degradation in the original manufacturing model as a random decrease in beta. Then, if all the manufactured units in this aging model pass the performance specification, this is a good indication that any single unit will pass specifications over time. Essentially, by changing the manufacturing model to represent aging, we are manufacturing "young," "middle-aged," and "old" units all at once. Also, we are requiring all units, no matter what their age, to pass specification testing. Optimization of simulated manufacturing yield allows the designer to reduce the effects of component aging even before a single unit is produced.

The same modeling philosophy also works for environmental effects, such as temperature. In this case we will model manufacturing as simultaneously producing both hot and cold units. Again, the designer can consider the effects of temperature on the unit performance in the initial design phase.

1.7 DESIGN

The initial design part of Figure 1.5 is the primary focus of this book. As we said previously, in the initial part of the design process, the engineer determines the unit structure, technology, and parameter nominal values which will best fit the given specifications. In this process there are many decisions and trade-offs to be made. For instance, the design mathematics may require an RC product of 10^{-6}. It is the designer's job to choose whether R should be 1 or 10 kΩ. This multitude of choices makes the design of any unit a process that can have several "correct" results. At times, making good choices from among all of the alternatives can be very difficult. To aid the designer in choosing the best unit-design parameters, optimization methods are sometimes used.

1.7.1 Single-Point Optimization Design Approach

Single-point optimization methods use sophisticated numerical techniques to determine the unit parameter values that provide the best performance as defined by an error function. The error-function formulation indicates the designer's preference for desired unit performance. Hopefully, the best performance is somewhat better than called for by the unit specifications (i.e., a conservative design). This way, if the performance degrades during manufacture, the units may still meet specifications. We call this approach *single-point* design.

The single-point approach to unit optimization is widely used. However, it can lead to designs that perform poorly in the manufacturing environment. Consider Figure 1.12 where unit performance is plotted versus a parameter value. Designers using single-point optimization will choose the parameter value given at point P_0, the point of maximal unit performance. In Figure 1.13, depiction of the single-point optimized parameter value P_0 is accompanied by the parameter's expected range of variation due to manufacturing. Also illustrated is the original specification for unit performance. Only performance higher than the specification

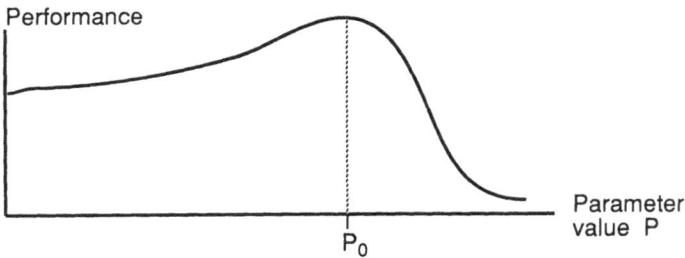

Figure 1.12 Performance versus a parameter value. The parameter value P_0 maximizes performance.

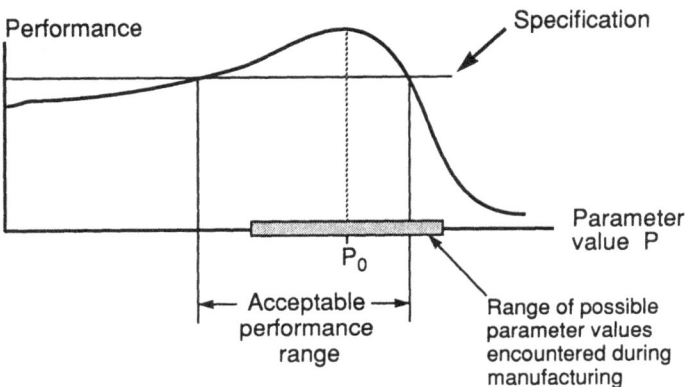

Figure 1.13 Performance versus a parameter value for an uncertain parameter. In this case, some fraction of the randomly produced parameter values cause unit failure.

is acceptable. Upon examination of Figure 1.13, we see that as the parameter value varies during manufacture the performance varies. When the parameter value is at its highest allowable extreme, the unit is out of specification, and it therefore fails the performance tests. Although excellent performance is achieved using the exact parameter value P_0, the design is not well-suited to the manufacturing environment. This is because unit failure occurs for all parameter values in the top portion of the potential range of values. The single-point optimization approach illustrates the concept of *brinksmanship* design [8].

1.7.2 Extending Single-Point Procedures With Statistical Design

The purpose of any statistical-optimization approach is to find not the nominal parameter value that will give the best performance, but a nominal parameter value that provides good yield during manufacture. Figure 1.14 shows the same performance curve as Figure 1.13, with a new nominal parameter P_0'. The variation on P_0' is the same as on P_0, with $P_0' < P_0$. Notice that P_0' is in the center of acceptable parameter values as given by the performance specification (i.e., design centering). When the unit is manufactured with the new parameter nominal point P_0', no parameter values will be encountered such that performance falls below specification.

The cost of using parameter value P_0' rather than P_0 in this design may be negligible, as both parameter values have the same variation. Yet using P_0' in the design will significantly increase the yield. This is the power and significance of using the statistical-design approach.

In the statistical-design approach, the designer investigates unit performance and optimizes it over the entire set of parameter values that will be encountered during manufacture. We believe that this "global" approach to design will always be beneficial to the practical success of a design.

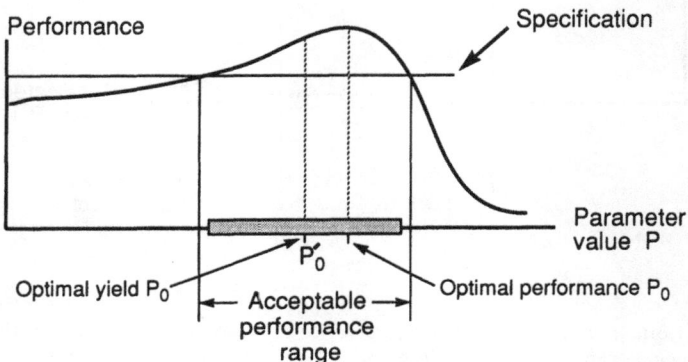

Figure 1.14 Performance versus a parameter value, and the parameter value P_0' which maximizes yield.

From this discussion it is apparent that the designer who wishes to practice statistical design will need accurate and reliable manufacturing models, including statistical descriptions of the range of parameters that the design will encounter during manufacture. Also, the designer will need accurate test models, including the specification windows used during unit testing. With all of this information to process, a computer will be needed. However, since it is commonplace to use a computer for single-point unit design, this requirement should not burden the designer. Figure 1.15 shows pictorially the requirements for extending the traditional single-point design approach.

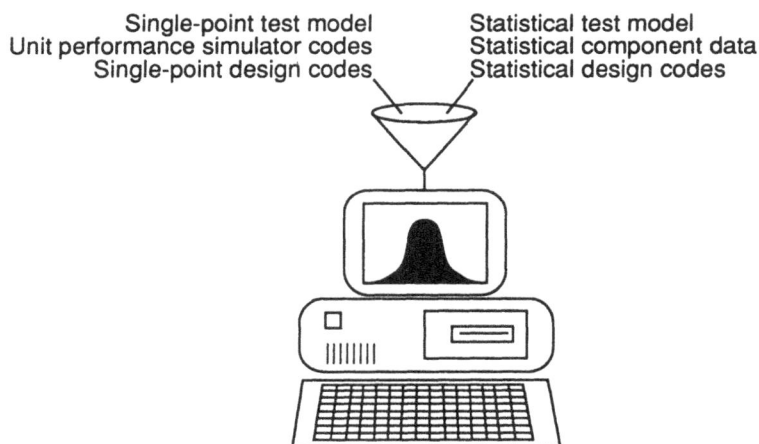

Figure 1.15 Requirements for extending the traditional single-point design method.

1.8 SOURCES OF PARAMETER VALUE UNCERTAINTY

From the point of view of a designer in the initial stages of unit design, there is a need to predict, as accurately as possible, the performance test results for a given manufactured unit. If this is possible, the designer can, with confidence, optimize a single unit's performance. However, there are uncertainties which degrade the designer's ability to know exactly how any single manufactured unit will perform. There are three major sources of uncertainty:

- Unit component modeling errors in the designer's simulator;
- Parameter-value uncertainty due to fabrication imprecision, environmental effects, and component aging;
- Testing error due to calibration uncertainty and measurement noise.

The logical flow and arrangement of these errors is illustrated in Figure 1.16.

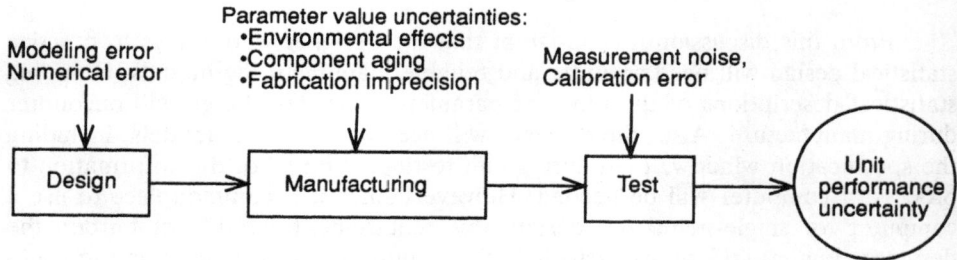

Figure 1.16 Sources of design-performance uncertainty.

A careful assessment of each of these error magnitudes can aid in understanding the process of design for manufacture and any potential limitations. Also it can help to avoid spending time and money in areas of the design process that will have little effect on the overall results. If one uncertainty dominates the overall process, reducing the others will have little effect on the final outcome. We feel that in many cases that parameter variation stemming from manufacturing imprecision will usually dominate the designer's uncertainty in predicting a manufactured unit's performance.

For instance, consider an example of the modeling, manufacturing and testing of a two-port described by S-parameters. Suppose that modeling errors in the designer's simulator are typically less than ± 0.01 in magnitude and $\pm 5°$ in phase. The measurement error using a modern network analyzer is typically on the order of ± 0.001 in magnitude and $\pm 1°$ in phase. However, the parameter variations in the manufacturing environment may cause magnitude variations of ± 0.2 and angle variations of $\pm 50°$. In this case, getting a more accurate simulator or test set will not help to design a more manufacturable unit.

Table 1.1 presents a hypothetical set of S-parameters predicted using two different simulators, one that models the two-port exactly, and one that has an

Table 1.1
Statistical Variation of Two-Port S-Parameters Using Two Simulators

	$|S11|$ using perfect simulator	$|S11|$ using erroneous simulator
unit 1	0.6153	0.6260
unit 2	0.5899	0.5821
unit 3	0.6056	0.6099
unit 4	0.6105	0.6088
unit 5	0.5923	0.5883
unit 6	0.5953	0.5982
unit 7	0.5804	0.5872
unit 8	0.5945	0.6041
unit 9	0.6022	0.6073
unit 10	0.6137	0.6213
$|S11|$ mean	0.600	0.603
$|S11|$ variance	0.011	0.014

error of ±0.01 or less. Each row in Table 1.1 lists the manufactured unit's predicted performance, given the parameter variations present during manufacture. A quick look at Table 1.1 reveals that the results in both columns, although differing by the simulator error, have essentially the same statistics. The simulation error shifts the mean by approximately 0.003, while the performance variance due primarily to the parameter-value uncertainty is larger at approximately 0.011. In essence, the simulator error has little effect on predicting the performance statistics because the parameter variation dominates the performance uncertainty.

There is little benefit to optimizing each step in the manufacturing process (design, manufacture, and testing) individually. Instead we need to look at the design process as a whole. A general rule is to compare the simulator performance error and the measurement error with the performance uncertainty due to the manufacturing variations. If the simulator performance and measurement variations are small compared to the manufacturing performance variation, the simulator and the test set are probably fine. In fact, a conservative approach to design would be to include the simulator and test uncertainties in the manufacturing statistical models. This approach to statistical design and all of its benefits has yet to be fully explored.

1.9 WHEN TO USE STATISTICAL CIRCUIT DESIGN

The examples in Section 1.10 indicate that there are often considerable benefits realized from statistical circuit design. But an important question arises: when is statistical design not necessary? Statistical design is not necessary when:

- A unit has essentially no parameter variation;
- The performance of the unit is not sensitive to uncontrollable parameters;
- The cost is considered less important than the highest possible performance.

There are many instances when these criteria are met. However, these criteria have an associated cost, as will be shown.

1.9.1 Voltage Divider

Figure 1.17 shows a passive voltage divider. Usually in this design, the resistor values can be tightly controlled; thus, the divider's performance can also be tightly

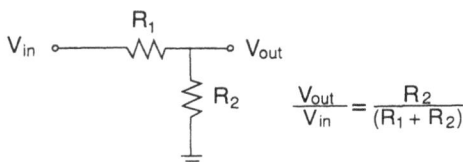

$$V_{in} \circ \!\!-\!\!\bigwedge\!\!\bigwedge\!\!-\!\!\circ V_{out} \qquad R_2 \qquad \frac{V_{out}}{V_{in}} = \frac{R_2}{(R_1 + R_2)}$$

Figure 1.17 A simple voltage-divider circuit.

controlled. If the two resistor values can be controlled during manufacture, the performance will not vary, and there will be no need for statistical design.

However, controlling the value of the resistors during manufacture usually has an associated cost—0.1% tolerance resistors are more expensive than 10% tolerance resistors. With proper statistical design, perhaps the same manufactured performance could be achieved with less costly components.

1.9.2 Low-Frequency Operational Amplifier

Figure 1.18 illustrates a low-frequency operational-amplifier circuit. The low-frequency gain of this circuit is closely approximated by R_2/R_1. This is independent of the operational-amplifier characteristics, given that the gain bandwidth product of the operational amplifier is much greater than the closed-loop gain bandwidth product specification. From a manufacturing viewpoint, this is ideal, because the active device parameters do not need to be tightly controlled. The gain is set and controlled by two resistor values, as in the previous passive-divider example. But this is the advantage of feedback—it desensitizes the response to the elements in the feedforward path. Feedback is often employed in electronics design to improve manufacturability. In a circuit like this, the designer may not need to use statistical design and analysis.

There is a price for this insensitive design: sacrificed performance. The feedback method of circuit desensitization will always trade performance for manufacturability. In this design, the operational amplifier has a gain bandwidth product that is probably one order of magnitude greater than the closed-loop manufacturable design. The question then becomes: "Can I afford to trade performance for manufacturability?"

In most "high-performance" applications or "state-of-the-art" technologies, a large surplus performance margin is nonexistent. Therefore the designer has no performance margin to trade for manufacturability. In these cases the designer is stuck with both the sensitivity inherent to the design and the parameter variability encountered during manufacture. Figure 1.19 illustrates this point with a typical lumped-component high-frequency amplifier.

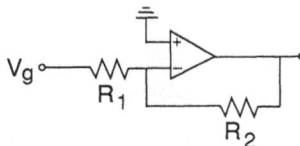

Figure 1.18 A low-frequency operational-amplifier circuit.

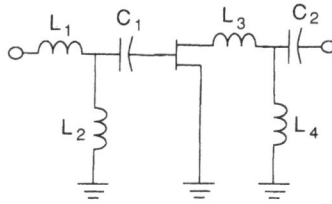

Figure 1.19 A typical high-frequency-amplifier circuit.

1.9.3 High-Frequency Amplifier

If we were to ask a designer to determine the gain of the amplifier in Figure 1.19, he would ask many questions about the transistor. The gain of this amplifier is critically dependent not only upon the gain of the transistor, but also on its input and output impedance and its bias point. From a manufacturing point of view this is unfortunate because variations in the active device parameters are usually the hardest to control during manufacture. Any variation here will directly cause variation in the performance of the amplifier. Does this mean that the design is faulty and should be discarded? Not necessarily. Usually, when a designer produces this kind of design, the goal is to achieve a large fraction of the transistor's maximum available gain. Feedback is not usable here because there is no excess gain to trade for manufacturability. The amplifier will inherently be sensitive to the active device's performance. In cases like this, statistical analysis and design are essential to the design and manufacturability of the unit.

In summary, the techniques used in this book are not always needed. There are some very fine design methods, notably feedback, which can be used to minimize performance sensitivity to uncontrollable parameters. However, where designs are pushed to their extremes, where specifications are set at the limits of the technology, and where performance is costly and requires careful management, performance sensitivity is a natural and unavoidable consequence. It is in these areas that a designer can fully leverage the power of statistical analysis and design techniques.

1.10 EXAMPLES OF STATISTICAL CIRCUIT DESIGN

In Section 1.7 we saw that when the performance versus parameter value curves are not symmetric, there is the possibility of a brinksmanship design. The example was simple and one-dimensional. In an academic sense, using a simple example is appropriate, but the inquisitive reader might ask if real units can exhibit brinksmanship behavior. This section briefly presents four examples of circuits and systems that demonstrate brinksmanship behavior. In addition, we will suggest how each example will benefit from the application of statistical design. We have found that most circuits and systems display some degree of brinksmanship behavior. These examples were not difficult to find, and are representative of what statistical design can offer. We should point out that statistical design is not a panacea for manufacturing problems. Rather, we believe that for the modest investment in the extension of current design methods, the design engineer can submit a design to production with a level of confidence otherwise unattainable.

1.10.1 Butterworth Filter

Figure 1.20 shows a third-order active Butterworth lowpass filter. The specification calls for a gain of 3 dB ± 0.5 dB at 4 kHz. The component statistics chosen for this example are $\pm 10\%$ uniform and independent variation on the capacitors and no variation on the resistors. The nominal parameter values are 3979.0 kΩ for the resistors and 0.01 μF for the capacitors. The yield for this circuit is approximately 30%. After design centering on the capacitor values, $C_1 = 0.0238$ μF, $C_2 = 0.0048$ μF, $C_3 = 0.0051$ μF, and the yield increases to 95%.

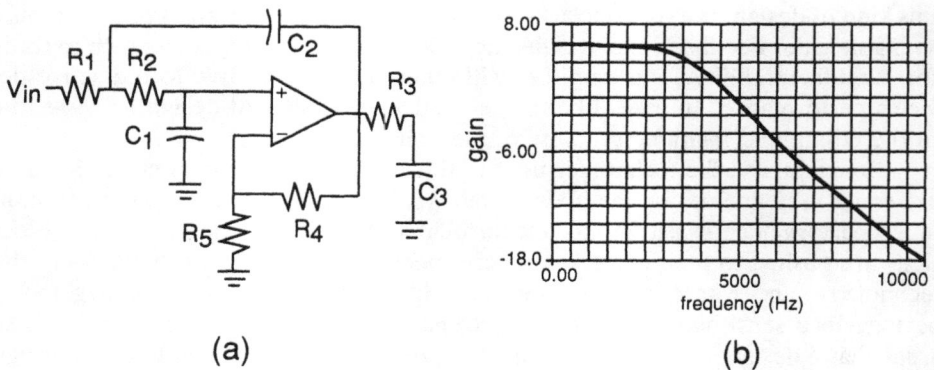

Figure 1.20 (a) Third-order Butterworth filter; (b) nominal parameter value frequency response.

1.10.2 A 2- to 6-GHz Feedback Amplifier

Figure 1.14 shows a feedback-amplifier design that might be implemented as a GaAs integrated amplifier. The specifications for this amplifier are S11 < −9 dB,

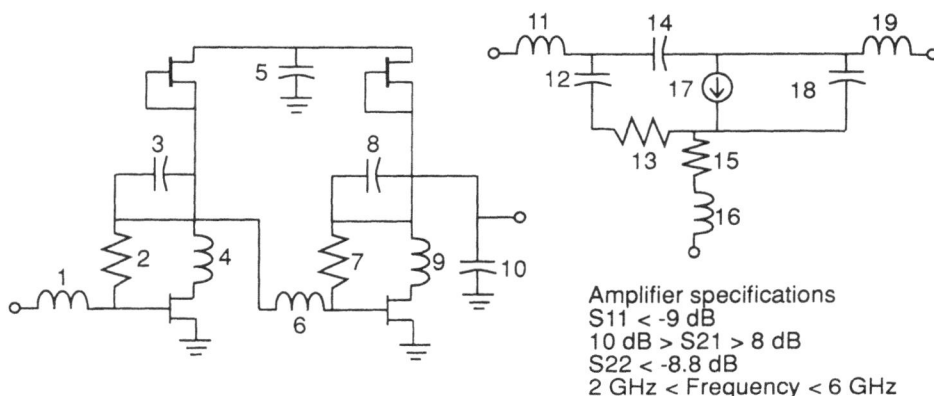

Amplifier specifications
S11 < -9 dB
10 dB > S21 > 8 dB
S22 < -8.8 dB
2 GHz < Frequency < 6 GHz

Figure 1.21 Feedback-amplifier configuration.

8 dB < S21 < 10 dB, and S22 < −8.8 dB over the frequency range from 2 to 6 GHz. The transistor model used is also shown in Figure 1.21. All the parameters in this circuit were given uniform ±10% independent statistical variations. The single-point optimized design gives a yield of approximately 29%, while the statistically designed circuit gives a yield of approximately 49%. Although this is a feedback design, performance sensitivity, especially to component 4, remains because the gain requirement does not allow enough feedback to adequately desensitize the circuit.

1.10.3 Satellite-Receiver System

Figure 1.22 shows a typical 4.0-GHz microwave-receiver system [9]. There are four statistical parameters in this system: g1 = 45 dB ±10%, a2 = −10 dB ±10%, lo1 = 1.7 GHz ±0.25%, and BW1 = 33 MHz ±10%, where g1 is the first amplifier gain, a2 is the conversion gain of the second mixer, lo1 is the second-mixer local-oscillator frequency, and BW1 is the system-output filter bandwidth. These parameters were chosen as independent and uniformly distributed for this example. The specifications for the system were output signal-to-noise ratio > 11 dB, and in-band group delay < 0.45 μsec. The initial system design has a yield of approximately 20.9%, while the yield-optimized system has a yield of 100%. These dramatic results are due to the fact that the single-point design, although giving good single-point performance, does not have the output carrier centered in the output filter. This imposed a sensitivity to the output-filter bandwidth which was detected and minimized using statistical design.

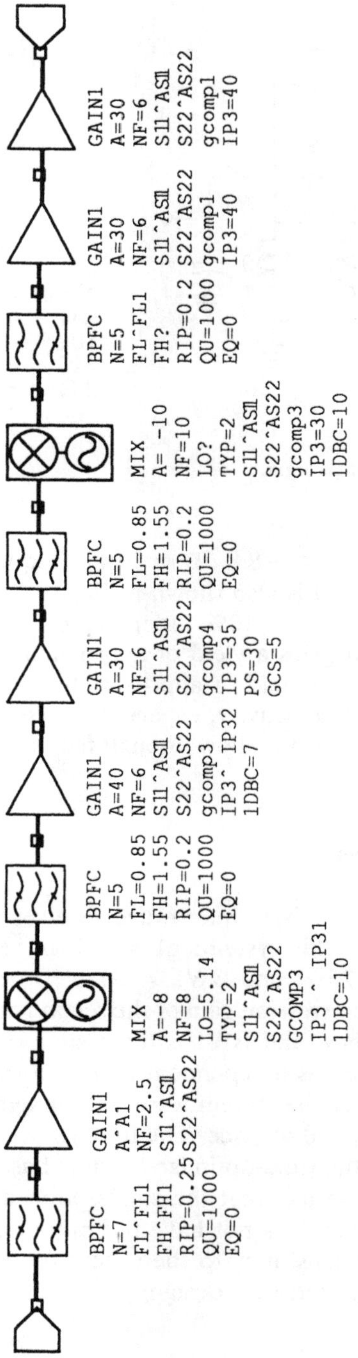

BPFC
N=7
FL^FL1
FH^FH1
RIP=0.25 S22^AS22
QU=1000
EQ=0

GAIN1
A^A1
NF=2.5
S11^AS11 S22^AS22

MIX
A=-8
NF=8
LO=5.14
TYP=2
S11^AS11
S22^AS22
GCOMP3
IP3^ IP31
1DBC=10

BPFC
N=5
FL=0.85
FH=1.55
RIP=0.2
QU=1000
EQ=0

GAIN1
A=40
NF=6
S11^AS11
S22^AS22
gcomp3
IP3^ IP32
1DBC=7

GAIN1
A=30
NF=6
S11^AS11
S22^AS22
gcomp4
IP3=35
PS=30
GCS=5

BPFC
N=5
FL=0.85 FH=1.55
RIP=0.2
QU=1000
EQ=0

MIX
A= -10
NF=10
LO?
TYP=2
S11^AS11
S22^AS22
gcomp3
IP3=30
1DBC=10

BPFC
N=5
FL^FL1
FH?
RIP=0.2 S22^AS22
QU=1000
EQ=0

GAIN1
A=30
NF=6
S11^AS11 S22^AS22
gcomp1
IP3=40

GAIN1
A=30
NF=6
S11^AS11 S22^AS22
gcomp1
IP3=40

Figure 1.22 4.0-GHz satellite-receiver block diagram.

1.10.4 Tunable Active Filter

This example is included to show the effects of statistical models on the results of statistical design—in this case, yield calculation. Figure 1.23 shows a tunable active filter [10]. The yield for this filter was analyzed using two different statistical parameter models. Initially the parameters were assumed to be independent with a given spread on each parameter. The yield is approximately 27%. Then the parameters were assumed to be correlated, with an optimum correlation developed in the reference, and using the same parameter variations as in the previous case. For the correlated parameter case the yield is approximately 95%. This increase in yield is solely due to a change in correlation of the parameters. This illustrates the importance of having accurate statistical models when performing yield calculation. Statistical design is also dependent on the parameter statistics (Chapter 6 focuses on this topic).

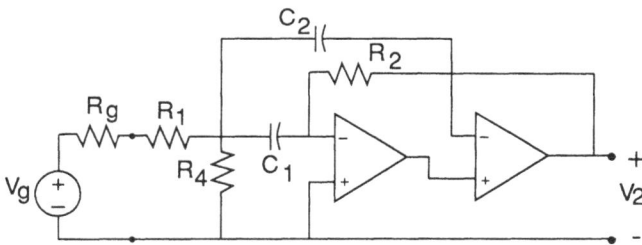

Figure 1.23 A tunable active filter.

1.10.5 Summary of Examples

The first three design examples show the power of statistical design for three different applications. The fourth example shows the need for accurate and reliable statistical models when performing statistical design. Most circuits and systems can be improved by the application of statistical-design techniques.

1.11 CONCLUSION

Robust design, quality design, and concurrent engineering are all topics of great interest today, with yield and reliability design being important factors in each. The models we introduce here are general enough that they should apply to most design and manufacturing situations. Present-day design practices that do not model or consider parameter variations can leave parameter sensitivities undetected until the unit is manufactured or released to the customer. On the other hand, statistical-design techniques like those introduced in this book do consider unit performance over the range of parameter values expected during manufacture and service life. Therefore, statistical design takes aim at reducing the sensitivity to critical parameter variations, and thus minimizing the chance of unwanted failures.

1.12 IMPORTANT IDEAS FROM CHAPTER 1

Section 1.1
- Design for yield and reliability will become economic necessities in the coming decade.
- Design for yield and reliability requires communication between all departments in a company, particularly between design, manufacturing, and testing.
- Design for yield and reliability is closely related to robust design.
- Quality is often defined as small performance variance during the manufacture and service life of the unit. A large part of design for quality lies in the area of statistical design.
- Robust design is the intelligent trade-off between, and optimization of, unit cost, reliability and manufacturability.

Section 1.2
- Manufactured products are described as units in this book. Units are made up of components, which are described in terms of parameters.

Section 1.3
- The goal of a successfully designed product is to pass performance tests after manufacture.

Section 1.4
- Three important questions that must be asked are:
  1. How do we design a manufacturable unit?
  2. How do we eliminate, or at least reduce, the costly iteration cycle between design and manufacture?
  3. How do we design in reliability?
- One of the central objectives of this book is to break down the "wall" between design and manufacture.

Section 1.5
- The work in this book addresses parametric unit failures as opposed to catastrophic ones.

Section 1.6

- There are three ways to increase unit yield:
  1. Tightly control the manufacturing parameter variations;
  2. Use a design structure where performance is insensitive to uncontrollable manufacturing parameters;
  3. Choose the nominal parameter values for the given design structure that are best suited to the particular manufacturing environment.
- Parameter aging and environmental effects on a unit can be included in our manufacturing model without difficulty.

Section 1.7

- Single-point optimization can design in unit-performance sensitivities that lead to brinksmanship designs. Therefore, we implore the reader to "ban the nom" (i.e. nominal), and to use single-point optimized design as the starting point for statistical-design procedures such as yield optimization.
- In statistical design, the designer investigates and optimizes unit performance over the entire set of parameter values that will be encountered during manufacture. Note that variations due to environmental effects and service life can also be included.

Section 1.8

- We feel that the manufacturing parameter variation will usually dominate the designer's uncertainty in predicting a manufactured unit's performance.

Section 1.9

- Statistical design is not necessary when:
  1. A unit has negligible parameter variation;
  2. The performance of the unit is not sensitive to uncontrollable parameters;
  3. Cost is considered less important than highest possible performance.
- In some situations, good design practice can reduce the need for statistical design.

Section 1.10

- The examples of a Butterworth filter, a feedback amplifier, and a satellite receiver all illustrate the general applicability of statistical design.
- The tunable-active-filter example illustrates the importance of the statistical models used when performing statistical analysis and design.

REFERENCES

[1] S.G. Shina, "Concurrent Engineering, New Rules for World Class Companies" *IEEE Spectrum,* Vol. 28, No. 7, July 1991, pp 22-26.

[2] R. Aguayo, *Dr. Deming, The American Who Taught the Japanese About Quality,* Carol Publishing Group, A Lyle Stewart Book, 1990.

[3] G. Taguchi, *Introduction to Quality Engineering, Designing Quality into Products and Processes,* Tokyo, Japan: Asian Productivity Organization, 1986.

[4] B. King, *Better Designs In Half the Time,* Methuen, MA: GOAL/QPC, 1989.

[5] K. Ishikawa, *Guide to Quality Control,* Tokyo, Japan: Asian Productivity Organization, 1990.

[6] D. Agnew, "Design Centering and Tolerancing Via Margin Sensitivity Minimization," *Proc. IEE,* Pt. G, Vol. 127, No. 6, 1980, pp. 270-277.

[7] J. Purviance and D. Monteith, "High-Yield Narrow-Band Matching Structures," *IEEE Transactions on Microwave Theory and Techniques,* Vol. 36, No. 12, Dec. 1988, pp. 1621-1628

[8] R. Brayton, R. Spence, *Sensitivity and Optimization,* New York, NY: Elsevier, 1980.

[9] R. Cook and J. Purviance, "Statistical Design for Microwave Systems," Proc. of the IEEE MTT-S Int. Microwave Symposium, Dallas, TX, June 1991, pp 679-682.

[10] H.L. Abdel-Malek, and S.O. Hanson, "The Ellipsoidal Technique for Design Centering and Region Approximation," *IEEE Trans.* on CAD, Vol. CAD-10, No. 8, 1991, pp. 1006-1014.

Chapter 2
Yield

"Better-designed, higher quality products with a shorter time to market means higher profits, and trouble-free product introductions often win market share away from competitors."[1]

2.1 INTRODUCTION

The goal of this book is to introduce methods that produce reliable, high-yield designs for microwave circuits and systems. Accordingly, one of those methods is to maximize the manufacturing yield for the unit that we are designing. Yield is defined as the ratio of manufactured units which pass performance testing to the total number of manufactured units (in the limit as the number of manufactured units approaches infinity). Also, remember we deal with only parametric yield (see Section 1.5.3). But before we set out to maximize yield (by writing or using yield-optimization computer-aided-design software), we must set the foundation by discussing yield and its properties. There are two broad ways to classify the yield-calculation problem—this chapter presents both of them.

2.2 TWO WAYS TO DESCRIBE YIELD

There are two different approaches helpful in describing and calculating yield. Each is useful for developing certain aspects of yield and its calculation. Existing techniques for yield optimization are usually best explained using one of the two approaches. Use of one of the two available viewpoints is similar in approach to solving systems problems in either the time domain or the frequency domain. In different design circumstances, one approach is usually preferred.

2.2.1 Mathematical Viewpoint: Calculating Yield

The mathematical approach to describing yield involves a multidimensional integral equation. Familiarity with the mathematical approach offers an understanding of the subtleties of yield calculation such as the requirements for practical calculation of yield with closed-form analytical techniques. Further, the mathematical viewpoint into yield calculation can be employed when studying any of the elegant analytic solutions available in the literature.

2.2.2 Geometric Viewpoint: Seeing Yield

The geometric approach gives the reader important visual insight into the yield calculation problem. We use two-dimensional geometric examples to illustrate the concepts because they are graphically simple. However, we hope that the reader can extrapolate the geometric approach on multidimensional problems (although multidimensional graphs quickly become impossible to draw). The authors prefer to use the geometric approach for describing yield and its calculation because of its visual, intuitive nature; therefore, we use the geometric approach whenever it is remotely convenient.

Section 2.3 introduces some necessary geometric underpinnings. Section 2.4 is a brief review of statistical concepts. Sections 2.5 and 2.6 present yield, first as a geometrical concept, and then as a mathematical integral.

2.3 PARAMETER SPACE AND PERFORMANCE SPACE

Yield can easily be viewed in terms of two spaces, *parameter space* and *performance space*. Any realization of a unit can be represented as a point in parameter space, and the unit's performance as a point in performance space. Together, these spaces comprise a valuable tool for the understanding of yield calculation and optimization.

2.3.1 Parameter Vector, P

Each unit, as it is manufactured, is represented by a collection of parameters specified by a designer before manufacturing began. Referring to the description of the manufacturing model in Chapter 1, the parameters are variables whose statistics are defined by the manufacturing environment, and are communicated to the designer via the manufacturing model. As an example, the resistive voltage divider shown in Figure 2.1 could be represented by two parameters, the resistance values of R_1 and R_2. During manufacture each divider is assigned different values for R_1 and R_2. Using the notation (R_1, R_2) to represent the unit parameters, realizations of the voltage divider for four units might be: (990,998), (1002,995), (999,1006), and (1004,1001). A more complicated unit has more parameters. Any unit has a total of n parameters, where n is an integer. For some circuits and systems, n could be in the hundreds.

Next, we denote the general parameter collection as P, where we will deal only with continuous numerical parameters such that P is always an ordered array of continuous numbers. P can also be thought of as a *vector*, where the ith element of P is given as p_i, for i = 1 to n. Therefore the parameter vector is expressed as:

$$P = (p_1, p_2, \ldots p_n) \tag{2.1}$$

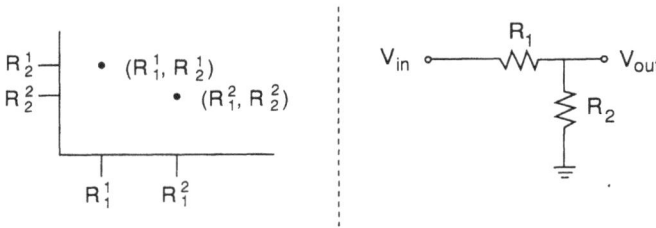

Figure 2.1 Parameter space, $\mathcal{P}$, for the resistive voltage divider. R_1 and $R_2 > 0$.

2.3.2 Parameter Space, $\mathcal{P}$

Each manufactured realization of a unit can be described in terms where the parameter vector has element values matching those of the realization. If we wish to characterize all the many different unit realizations possible during manufacture, we only need to determine how many different parameter vectors are possible. In a formal way, we are assuming there is a one-to-one mapping between each manufactured unit and its parameter vector, P.

The collection of all possible parameter vectors defines a geometric region in n-space referred to as *parameter space,* which we denote as $\mathcal{P}$. If working with an n-dimensional space becomes troublesome, it is helpful to remember that each "point" in an n-space is actually a collection of n-numbers or a vector, rather than a single number. For the case of the resistive voltage divider there are two parameters, (R_1, R_2); the parameter space is therefore two-dimensional. The geometry of this two-dimensional space is graphed in Figure 2.1. Note that every point in this space represents a possible manufactured realization of the of the voltage-divider unit with parameter values (R_1, R_2).

If the manufacturing environment were totally unpredictable, we could (at least conceptually), perform a dart-throwing experiment in parameter space to model manufacturing. To do this, we would tack a graph of parameter space on the wall and throw a dart at the graph. The dart would land at a random point in parameter space that would represent randomly picked resistor values for a man-ufactured divider network. Because conceptual models like the dart board are very useful, we will use them throughout our study of statistical circuit and system design.

2.3.3 The Performance Space $\mathcal{M}$, and the Measurement Vector M

As $\mathcal{P}$ is linked to the manufacturing model, we develop a space linked to the testing model, called the performance space, $\mathcal{M}$. For each unit there is a measurement plan which entails the measurement and documentation of the unit's performance. In the general case, there are m measurements needed to verify the unit's post-production performance. We assume that each measurement is a continuous num-ber. By collecting the m measurement results into an ordered set $(m_1, m_2, \ldots m_m)$, we have the *measurement vector* = M. Examples of measurements might be gain, signal-to-noise ratio, noise figure, bandwidth, and reflection coefficients. Multiple frequencies are included by making a measurement at each frequency of interest. The collection of all measurements would completely describe a manufactured unit's performance.

When considering the collection of all numerical combinations of measure-ment vectors, some might be possible and some physically impossible. However, the entire collection defines an m-dimensional space which we call *performance space, $\mathcal{M}$*. For the formal definition of performance space, the vector M = $(0,0,\ldots0)$ must be present. Even if a manufactured unit cannot obtain each of the perfor-mances represented in this space, all numerical possibilities are nonetheless present in $\mathcal{M}$. A subregion of performance space will contain the performance possibilities for any unit design and its manufacturing environment. The performance space for the resistive voltage divider from the previous Section is illustrated in Figure 2.2. The two performance measurements for this circuit are assumed to be input resistance and voltage output.

If the unit performance is totally unpredictable, we could model the mea-surement process as another experiment throwing darts on to performance space, with each dart landing on a random point in $\mathcal{M}$. This is not a likely model however,

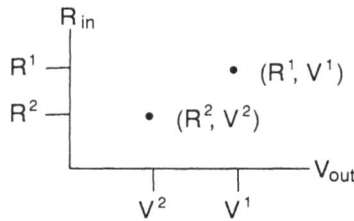

Figure 2.2 Performance space $\mathcal{M}$ for the resistive voltage divider.

as unit performance is usually confined to a region with fixed shape and not over the entire performance space.

2.3.4 Design Specification, S

Associated with each measurement in $\mathcal{M}$ is a design specification, S, for that measurement. A specification sets the acceptable limits or intervals for the respective measurements. Examples of typical specifications are: 10 dB < signal/noise; 25 dB < gain < 28 dB; 900 MHz < bandwidth < 1100 MHz. Note that each specification is defined by an inequality, sometimes *single-sided* and sometimes *double-sided,* that forms an acceptance interval.

It is not appropriate to have a specification established only through an equality, because no unit can maintain a single specification value during manufacture. For example, a specification of "gain = 25 dB" is an impossible specification to meet. No manufactured unit can obtain and preserve a gain of exactly 25 dB. Therefore, the specification must have the form "24.5 dB < gain < 25.5 dB." This inequality sets a performance window that is (hopefully) realistic for a manufactured unit.

In statistical design, we model the entire measurement and testing process as comparing each manufactured unit's measured performance with its associated performance specification. If any performance measurement is outside the specification interval, the unit will fail the test and be deemed unusable. Only if all measurements pass (i.e., fall within their respective specification intervals) will the unit pass the postmanufacture tests.

The designer must carefully collect specifications before the design process begins. The specifications represent the performance target for finished, fabricated units, and are essential to the initial design process. Often, specifications are set by the customer or end user. Complete specification data, including the intervals of acceptable performance for the unit, are key pieces of information needed by the designer before practicing statistical design and analysis.

2.3.5 Acceptable Performance Region, $\mathcal{M}_a$

The collection of m inequalities, which represent the unit specifications, describe the region in performance space called the acceptable performance region or *acceptability region, $\mathcal{M}_a$*. Essentially, $\mathcal{M}_a$ is the collection of all possible performance points in $\mathcal{M}$ that fall within specification. The $\mathcal{M}_a$ for the voltage-divider example is given in Figure 2.3. The size and shape of $\mathcal{M}_a$ is directly derived from the performance specifications, and for this example it is easy to determine its geometric properties such as its volume and boundaries. If we continue with the dart-throwing experiments and simulate unit performance by throwing darts at performance space, those darts falling within the acceptability region represent manufactured units that will pass all performance specifications.

The geometric model now consists of parameter space, $\mathcal{P}$, and performance space, $\mathcal{M}$. The last element of the model is the link between the two. This is accomplished by the performance function G(P).

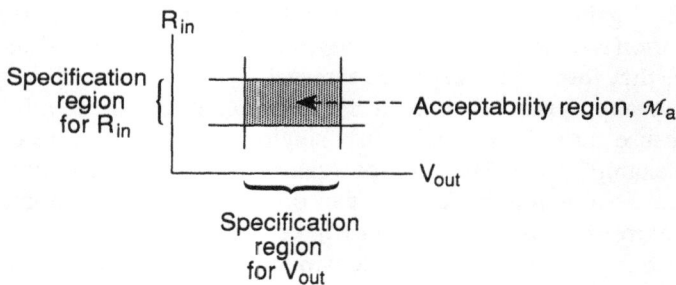

Figure 2.3 Acceptability region $\mathcal{M}_a$ in performance space for the resistive voltage divider.

2.3.6 Performance Function G(P)

The link between unit parameters and unit performance is usually clear. Unit performance is often determined by using some sort of simulation software. As unit parameters change, unit performance usually changes too. In fact, for each set of parameters, there is a unique performance. We formalize this relationship with the performance function, G(P). G(P) can be thought of as a mapping from parameter space (n dimensions) to performance space (m dimensions). This is stated as $G(P) = M$, or

$$\mathcal{P} \Rightarrow \mathcal{M} \qquad (2.2)$$

where P is the parameter vector, M is the measurement vector, and the arrow ($\Rightarrow$) indicates a mapping by the function G between the two spaces. If we have a good unit simulator and good component models, the simulator defines the mapping $G(P)$, because the simulator takes the unit parameters, P, and calculates the unit performance, M.

We find it useful to investigate the properties of this mapping of $G(P)$. First of all, $G(P)$ is usually nonlinear in terms of the parameters. Doubling a parameter value (e.g., substituting $2R_1$ in for R_1) generally will not double the unit's performance, even for linear circuits and systems.

The mapping of $G(P)$ is usually 1:1, but not "onto." In other words, each individual P maps into a unique M (1:1), but many different Ps can map into the same M (not "onto"). Thus there exists a unique M such that $G(P) = M$ for each P in parameter space, $\mathcal{P}$. However, $G(P)$ is not invertible; that is, there does not exist a unique P such that $G^{-1}(M) = P$ for each M in $\mathcal{M}$. A fundamental problem associated with all unit design lies in the noninvertibility of $G(P)$. If $G(P)$ were invertible, design could be accomplished simply by determining $G^{-1}(M)$ for the desired performance M. However, at our present level of understanding, only $G(P)$ is available for use in design, not its inverse. The mapping of parameter space to performance space is indicated in Figure 2.4.

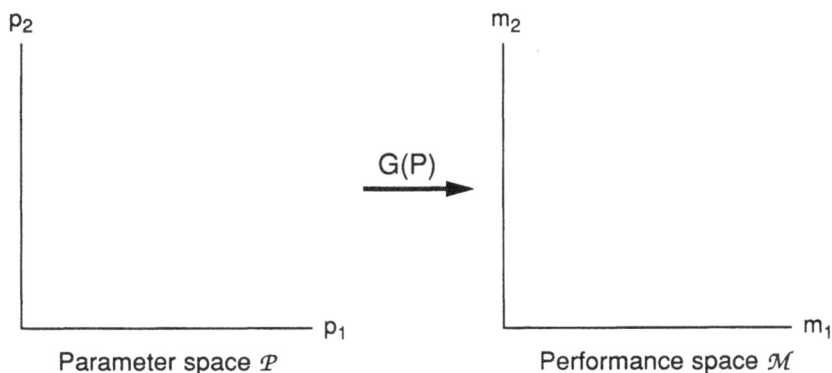

Figure 2.4 Mapping of parameter space into performance space by the performance function $G(P)$ (a two-dimensional example).

2.3.7 Acceptability Region in Parameter Space, $\mathcal{P}_a$

We now have a complete description of the acceptable unit performance in performance space, $\mathcal{M}_a$, as defined by the specifications. Because the net result of unit design is the unit parameter values, the real work of design lies in parameter space, $\mathcal{P}$, not performance space, $\mathcal{M}$. In most designs it is very helpful to know the region of acceptable parameter values in parameter space—the collection of parameter values which all give performance within specifications. Conceptually, this is easy to obtain. Consider applying G(P) to each point, P_i, in parameter space. We then tag each point with a label—pass or fail. The parameter vector P_i is tagged pass if $G(P_i)$ is within $\mathcal{M}_a$. Conversely, tag the parameter vector P_i with fail if $G(P_i)$ is not within $\mathcal{M}_a$. After tagging each point in $\mathcal{P}$ with a pass or fail, we collect all the pass points together into the set of acceptable parameter values. This set forms a region in parameter space called the *parameter acceptability region, $\mathcal{P}_a$*. This collection of parameter values forms the set of "good" parameters. These should be used in manufacturing because they provide acceptable performance and allow the unit to pass postmanufacture testing. One should note that since there are an infinite number of points in $\mathcal{P}$, the above stated procedure for determining $\mathcal{P}_a$ is computationally impractical. Nonetheless, this conceptual model for $\mathcal{P}_a$ is quite useful for both understanding and performing yield optimization. Consider Figure 2.5 which depicts parameter space, performance space, and their respective acceptability regions.

Again, we would like to emphasize that $\mathcal{P}_a$ is not directly calculable from the specifications, as it is usually impossible to obtain a closed-form relation for the unique transformation of points in measurement space to points in parameter space. Furthermore, $G^{-1}(M)$ is not generally available.

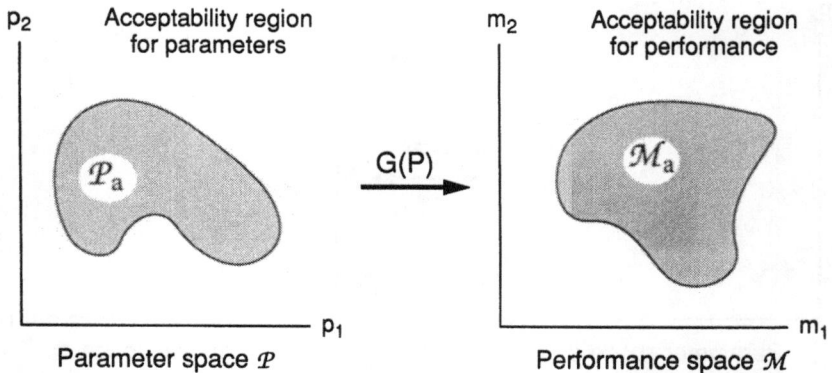

Figure 2.5 Mapping of the parameter acceptability region into the performance acceptability region by the performance function G(P).

Another important characteristic of $\mathcal{P}_a$ is that its shape is not "well-behaved;" it is typically neither convex nor connected (see Section 3.5.2), and as we shall see in Chapter 3, these properties make approximating $\mathcal{P}_a$ very difficult.

2.3.8 Example—A Voltage Divider

To summarize the geometric principles discussed thus far in a circuits context, consider the parameter space and performance space for the voltage divider described earlier. The divider is made from two resistors with nominal values of 35Ω $\pm 5\Omega$, and vary randomly according to a uniform and independent parameter distribution. The performance specifications are $50\Omega < R_{in} < 70\Omega$ and $4V < V_{out}$ $< 5V$. The parameter space and performance space for this circuit are shown in Figure 2.6. (The interested reader may wish to attempt the solution of $G^{-1}(M)$ for this divider example.)

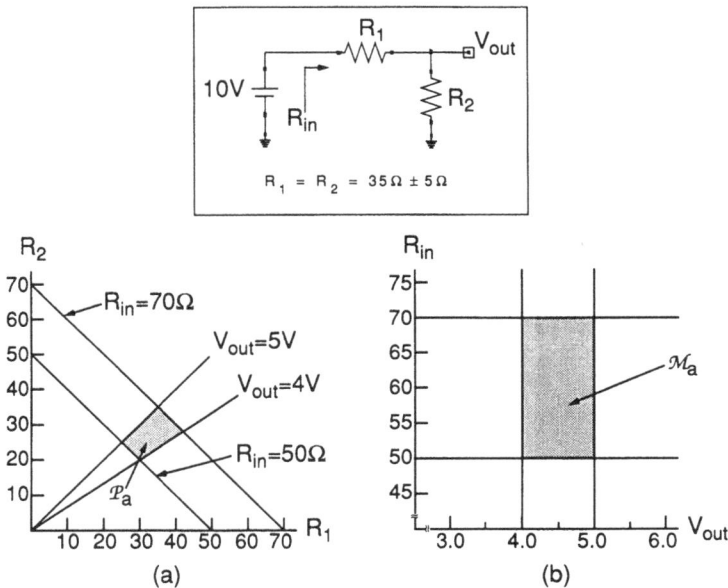

Figure 2.6 Parameter space and (a) performance space (b) acceptable regions for a voltage divider.

2.3.9 Tolerance Region, $\mathcal{T}$

Another concept useful to the geometric approach is that of tolerance, T. If a parameter's variation during manufacture is limited to a fixed range, this variation range is called the parameter's tolerance. For instance, if a 1-kΩ resistor value is guaranteed to lie within the range of 0.9 kΩ to 1.1 kΩ during manufacture, its tolerance is ± 0.1 kΩ. Often the tolerance is described as a percentage of the nominal value—for this resistor, T = $\pm 10\%$.

If each parameter is given a tolerance interval, the union of all these intervals in parameter space creates an n-dimensional region called the *tolerance region, T*. This is shown graphically in Figure 2.7. The tolerance region is important to statistical design because all of the parameter values encountered during manufacture are contained within *T*. So, if we wish to simulate the manufacturing environment, we can limit our choice of parameter values to those which lie within the tolerance region.

If certain parameters have an infinite extent, their tolerance interval is infinite. This is true for Gaussian distributed parameters. However a tolerance interval is usually assigned for each parameter like ± 2 or $\pm 4\sigma$, where σ is the parameter's standard deviation. If chosen carefully, limiting the extent of a Gaussian distributed parameter to a tolerance region can have surprisingly little effect on design centering and yield prediction. There are some analytic methods for yield calculation and optimization that require smooth parameter-distribution functions, and this truncation method may not be appropriate in those cases.

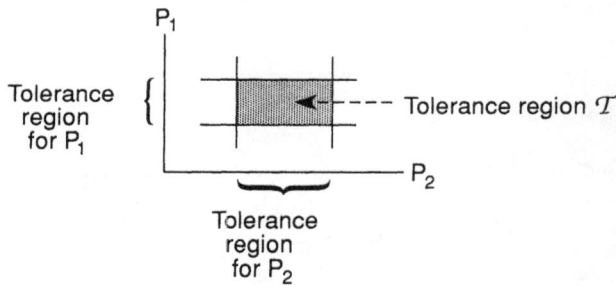

Figure 2.7 Tolerance region in parameter space.

2.4 PARAMETER STATISTICS

For any unit, its parameters are the key link between the unit's design and its measured performance. As discussed in Chapter 1, all parameters are, out of necessity, random variables because of the nature of manufacturing processes. A parameter that undergoes minimal variation during manufacture (actually a special case of a random variable) is referred to as a constant. To effectively model the manufacturing environment we must not only mathematically describe the design parameters' average or nominal values, but also properly represent their joint statistical variations. This requires the mathematics of probability. Our presentation of probability will be brief, but it contains the necessary elements for understanding statistical design. We have not placed this material in an appendix, as is often done, because it contains some uncommon warnings and examples important for proper understanding and practice of statistical design.

2.4.1 Random Variables

Due to the nature of the manufacturing process, the unit parameters are not fixed but vary randomly about some average or "nominal" value. The best way to model these parameters is to model them as random variables. The concept of a random variable is much more general than presented here, but this presentation is sufficient for the purposes of understanding statistical design.

A comment on notation: we use script variables to denote random variables. For instance, $\mathcal{P}$ denotes a random parameter or parameter vector, (i.e., a vector whose elements are random variables). P denotes a deterministic, or fixed, parameter or a parameter vector. In statistical design, it is important to keep track of the random and nonrandom components of the problem. This notation will make it clear whether we are talking about random or deterministic variables.

Associated with any random variable is an experiment or trial whose outcome is unknown prior to the exercise of the experiment. We call this a random experiment. For our purposes we assume that the experimental outcome is a continuous number. This random experiment and its outcome ($\mathcal{P}$) are shown descriptively in Figure 2.8.

The random experiment depicted in Figure 2.8 has an experiment-execution button. Every time the button is pressed, a numerical outcome is output from the experiment box. An example of a random experiment is rolling a six-sided die. The experiment execution is throwing the die, and the numerical outcome is the number showing on the top face of the die. Another random experiment might be going to a thermometer and accurately recording the room temperature. The room temperature then becomes the numerical outcome of the experiment.

In all cases, the result of the experiment is different every time the experiment is exercised, and the experimental outcome is impossible to predict exactly before

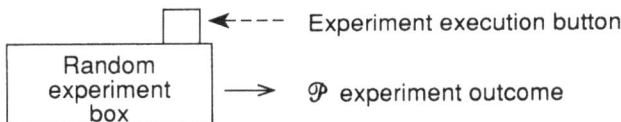

Figure 2.8 Random experiment and its outcome, $\mathcal{P}$.

the experiment execution. This uncertainty is modeled by assigning a random variable, $\mathcal{P}$, to the numerical outcome. The particular outcomes or groups of outcomes that result from the experiment are called the *events* or *samples* of $\mathcal{P}$, and are denoted with subscripted variables, with the subscript denoting the event, or trial number. We use roman notation for the event variables because once the trial has been executed, the trial outcomes are fixed numbers (for example, $P_1 = 5$, $P_{26} = 2.347893$, and $P_5 = 0.590385035943058$).

To mathematically represent the properties of $\mathcal{P}$, we calculate the probability that $\mathcal{P}$ will equal a certain number or range of numbers. An event E is a given grouping of outcomes. Examples of events are $\mathcal{P} = 5$ (a single outcome), and $2 < \mathcal{P} < 4$ (a range of outcomes).

The probability of an event, E, is defined as

$$\text{Probability(E)} = \text{Prob(E)} = \frac{\text{number of times E occurs}}{\text{total number of experiments}} \qquad (2.3)$$

in the limit as the total number of experiments approaches infinity.

This is called the *frequency-of-occurrence* definition of probability. Prob(E) is a number between zero and one. If Prob(E) = 0, the event E never occurs. If Prob(E) = 1, the event E always occurs. If Prob(E) = 0.5, the event occurs half of the times the experiment is executed.

This probabilistic model fits the parameter variations inherent in any manufacturing process and, in particular, the parameter variations in our manufacturing model. In this case, the random variable is the parameter value used in manufacture, and the random experiment is choosing the parameter from a collection of possible values, (i.e., from a parameter bin). For example, a 1-kΩ $\pm 10\%$ resistor can assume the values 0.9 kΩ < R < 1.1 kΩ. A result of the random experiment will be a realization of a particular resistor value such as 0.97342 kΩ. The resistor value, R, should really be treated as a random variable, $\mathcal{R}$, during manufacture.

2.4.2 Probability Density Function, $f_{\mathcal{P}}(P)$

The entire mathematical description of a single random parameter, $\mathcal{P}$, is contained in its *probability density function* (PDF), f(P). Again, the notation needs explanation. The PDF associated with the random parameter $\mathcal{P}$ should be denoted $f_{\mathcal{P}}(P)$, where the subscript $\mathcal{P}$ denotes the variable that f describes, and P is just a dummy variable in the argument of f(.). However, it is customary sometimes not to give the subscripted $\mathcal{P}$, and to use the dummy-argument variable to designate which variable f describes. For example, if there are two random parameters $\mathcal{P}_1$ and $\mathcal{P}_2$, each will have its own PDF, ideally denoted as $f_{\mathcal{P}1}(P)$ and $f_{\mathcal{P}2}(P)$ (remember, P is just a dummy argument variable). However, we will sometimes use the notation that $f_{\mathcal{P}1}(P) = f(P_1)$, and $f_{\mathcal{P}2}(P) = f(P_2)$. We are giving extra significance to the form of the dummy-argument variable for notational simplicity. If there is a case when this practice becomes ambiguous, we will use the more formal notation.

There are several mathematical definitions of the PDF, but the one that is most useful to us here is

$$\text{Probability } (a < \mathcal{P} < b) = \int_a^b f_{\mathcal{P}}(P)dP \qquad (2.4)$$

which indicates that the area under the PDF over an interval equals the probability that the variable lies on that interval. A graph of the PDF versus its independent variable gives much insight into the statistical behavior of the random parameter. Where the PDF is relatively large over an interval, the parameter is likely to assume the numbers in the interval. Conversely, where the PDF is relatively small over an interval, the parameter is unlikely to assume the numbers in the interval, but these numbers may still occur. Examples of PDFs include the Gaussian, the uniform, and the bimodal PDF, as illustrated in Figure 2.9. A consequence of the above definition of the PDF ((2.3) and (2.4)) is that the entire area under a PDF equals 1.0.

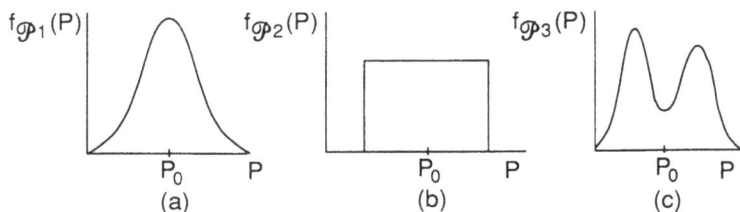

Figure 2.9 (a) Gaussian, (b) uniform, and (c) bimodal probability density functions. P_0 is the average or nominal value.

2.4.3 Average or Nominal Value

A useful number that helps summarize the properties of a random parameter is its average or *nominal* value (also referred to as the *mean* value). This number is the average value that the parameter assumes during manufacture. The nominal value of the parameters shown in Figure 2.9 is denoted as P_0. As illustrated in the graph of the bimodal PDF (Figure 2.9c), the nominal value is not necessarily the most probable value. In fact, for certain distributions, like bimodal or discrete distributions, it may be impossible for the average value to occur. The average value is given mathematically as

$$E(\mathcal{P}) = P_0 = \int_{-\infty}^{\infty} f_{\mathcal{P}}(P)dP \qquad (2.5)$$

where $E(.)$ is the expectation operation defined by the given integral. Close inspection of (2.5) should remind the reader of the *centroid* from physics. If we were to distribute gold ingots atop a wooden support in a way proportional to the bimodal distribution given in Figure 2.9, we would find the balance point at P_0 (Figure 2.10). Thus the average value can be described nicely as the center of gravity of the PDF. This helps us explain why the average value for a bimodal PDF rarely occur.

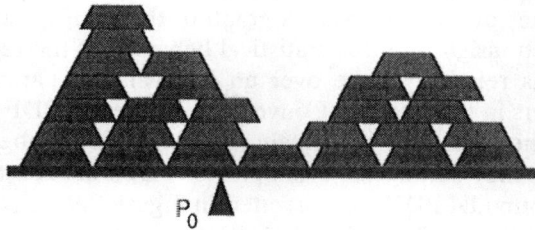

Figure 2.10 Balance point P_0 of a bimodal distribution of gold ingots.

2.4.4 Variance

A measure of the variation of a random parameter is given by its *variance*, σ^2. The variance is often likened to the amount of randomness (or energy) in a random parameter. Random parameters with high variance are characterized by a wide spread of possible values. Shown in Figure 2.11(a, b) are two uniform random-parameter PDFs. $f(P_2)$ has a wider range of possibilities (tolerance) than does $f(P_1)$. Thus, the variance of parameter 2 is larger than the variance of parameter 1. A constant parameter is modeled as a parameter with zero variance. The PDF of a constant parameter is the impulse function depicted in Figure 2.11(c). Variance is mathematically given by:

$$E(\mathscr{P} - P_0)^2 = \int_{-\infty}^{\infty} (P - P_0)^2 f_{\mathscr{P}}(P) dP \qquad (2.6)$$

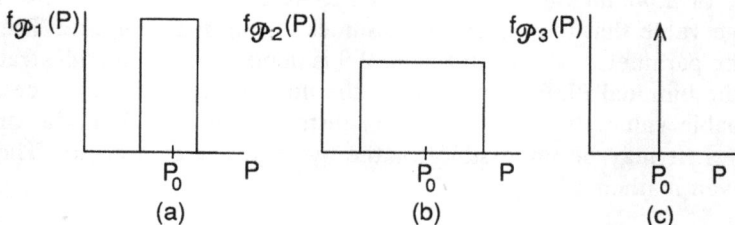

Figure 2.11 Uniform density functions (a) and (b), where the variance of (b) > variance of (a); density function of a constant random parameter (c). P_0 is the average or nominal value.

2.4.5 Higher-Order Moments

In general, there is a statistical quantity called the *nth-order moment* of the random parameter $\mathscr{P}$, given by

$$\text{nth-order moment} = E(\mathscr{P} - P_0)^n = \int_{-\infty}^{\infty} (P - P_0)^n f_{\mathscr{P}}(P)dP \qquad (2.7)$$

The variance is the second-order moment of a random parameter. In general, all higher order moments can be determined, if they exist.

2.4.6 Uniqueness

The mean and variance are useful in "summarizing" the statistical properties of a random parameter, but these two numbers do not uniquely describe the statistics of the random parameter. Two random parameters can have the same mean and the same variance but still not be statistically identical. Two parameters are *statistically equivalent* if they are described by the same probability density function. For instance, it is possible for a Gaussian random parameter and a uniform random parameter to have the same mean and variance; yet, the extent of the Gaussian parameter is infinite, and the extent of the uniform parameter is finite. We say that two random parameters are statistically equivalent only if they have the same PDFs (i.e., $\mathscr{P}_1$ and $\mathscr{P}_2$ are statistically identical if and only if $f_{\mathscr{P}_1}(P) = f_{\mathscr{P}_2}(P)$ for all P).

A necessary condition for statistical equivalence between two random variables is that their moments be equivalent for all orders. However, even if all the moments exist and are identical, the random parameters may not be statistically equivalent [2]. Many useful tests have been developed to examine whether samples from two random variables have been taken from the same PDF [3]. These are expounded upon in Chapter 6.

2.4.7 Multiple Random Parameters and Their Joint PDF

When two or more random parameters occur simultaneously from an experiment, they are called *jointly* distributed random parameters. Jointly distributed random parameters can occur in many ways. For instance if we define the experiment as throwing two dice at one time, and assign two random parameters, one for each die's outcome, these random parameters are jointly distributed. If the experiment measures simultaneously the temperature and the barometric pressure at a particular place in a room, the results are two jointly distributed random parameters. A more pertinent example is to let the experiment measure two transistor parameters, say C_{gs} and β, for a given manufactured transistor. There is no reason to limit the experiment to two simultaneous outcomes, except for simplicity of presentation. Jointly distributed random parameters are completely described by their joint probability density function $f_{\mathcal{P}_1\mathcal{P}_2}(P_1,P_2) = f(P_1,P_2)$, shown here for two random parameters. The joint density has the property that

$$\text{Probability } (a_1 < \mathcal{P}_1 < b_1, \text{ and, } a_2 < \mathcal{P}_2 < b_2) = \int_{-a_2}^{b_2}\int_{-a_1}^{b_1} f(P_1,P_2)dP_1dP_2 \quad (2.8)$$

If $f(P_1)$ is the density for $\mathcal{P}_1$, and $f(P_2)$ is the density for $\mathcal{P}_2$ then they are derived from their joint density by:

$$f(P_1) = \int_{-\infty}^{\infty} f(P_1,P_2)dP_2 \quad \text{and} \quad f(P_2) = \int_{-\infty}^{\infty} f(P_1,P_2)dP_1 \quad (2.9)$$

In this context we say that $f(P_1)$ and $f(P_2)$ are marginal densities for $\mathcal{P}_1$ and $\mathcal{P}_2$ derived from the joint density $f(P_1,P_2)$. A picture of a two-dimensional joint density with its related marginal densities is illustrated in Figure 2.12.

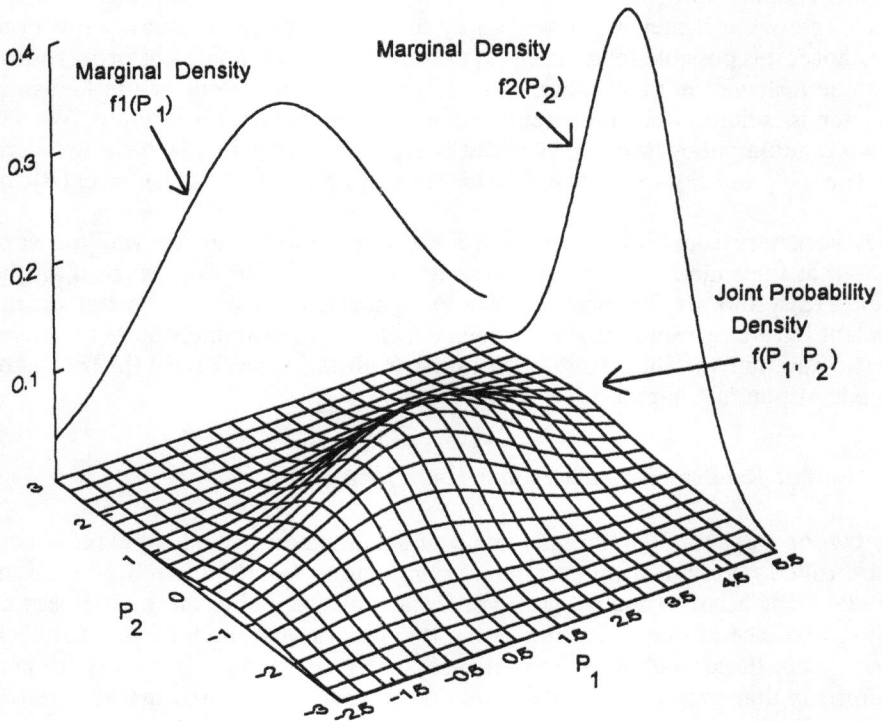

Figure 2.12 Joint density function, $f(P_1,P_2)$, showing the marginal density for $\mathcal{P}_1$ and $\mathcal{P}_2$, $f(P_1)$ and $f(P_2)$.

2.4.8 Covariance Matrix

The *covariance* of two random parameters is defined as:

$$E\{(\mathcal{P}_1 - P_{10})(\mathcal{P}_2 - P_{20})\} = \int_{-\infty}^{\infty}\int_{-\infty}^{\infty}\{(P_1 - P_{10})(P_2 - P_{20})\}f(P_1,P_2)dP_1dP_2$$
$$= COV\ (\mathcal{P}_1,\mathcal{P}_2) \qquad (2.10)$$

where again E(.) is the expectation operation and P_{i0} is the mean of $\mathcal{P}_i$. Let σ equal the square root of the variance, called the *standard deviation*. The covariance is often normalized by σ to become the correlation coefficient, R:

$$R_{12} = COV(\mathcal{P}_1,\mathcal{P}_2)/(\sigma_{P_1}\sigma_{P_2}) \qquad (2.11)$$

where $-1 < R_{12} < 1$.

If $R_{12} = -1$, then $\mathcal{P}_1 = -\mathcal{P}_2$. If $R_{12} = 1$, then $\mathcal{P}_1 = \mathcal{P}_2$. The larger the magnitude of R_{12}, the more statistically similar are the two random parameters, $\mathcal{P}_1$ and $\mathcal{P}_2$. If $R_{12} = 0$, $\mathcal{P}_1$ and $\mathcal{P}_2$ are statistically uncorrelated.

If there are more than two random parameters in any unit, the correlation among any pair of these parameters is conveniently represented in matrix form by the correlation matrix, C. For three parameters, C is given as

$$C = \begin{vmatrix} R_{11} & R_{12} & R_{13} \\ R_{21} & R_{22} & R_{23} \\ R_{31} & R_{32} & R_{33} \end{vmatrix}$$

By the properties of the correlation coefficient for this matrix, $R_{ii} = 1.0$ and $R_{ij} = R_{ji}$ for any integers i and j.

Figure 2.13 shows *scatter plots* for samples of four different joint distributions for random parameters, $\mathcal{P}_i$ and $\mathcal{P}_j$, for $i,j = 1,2; 3,4; 5,6;$ and $7,8$. A scatter plot is a graphical representation of samples for a pair of random parameters, where the pair form the x,y coordinates for the plot. In Figure 2.13, (a) shows a scatter plot for two uncorrelated and independent random parameters, $R_{12} = 0.034$; (b) shows a positive correlation, $R_{34} = 0.953$; (c) shows a negative correlation, $R_{56} = -0.951$; and (d) shows no correlation, $R_{78} = 0.013$ although clearly $\mathcal{P}_7$ and $\mathcal{P}_8$ are related to one another. This is because the correlation coefficient assumes a pair-wise linear relationship between variables. (Note that correlations are estimated from the 900 points shown in the figures.)

2.4.9 Independent and Uncorrelated Random Parameters

Two random parameters are called independent if their joint PDF can be separated into the product of two expressions: one expression is a function of one unique parameter value, and the other expression is a function of a second unique pa-

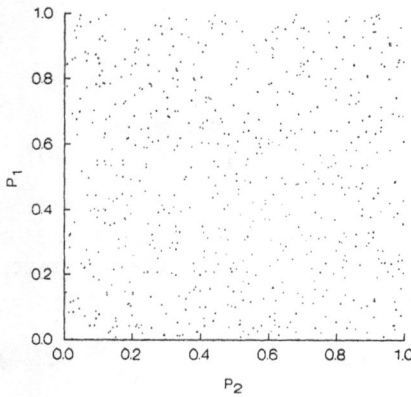

(a)

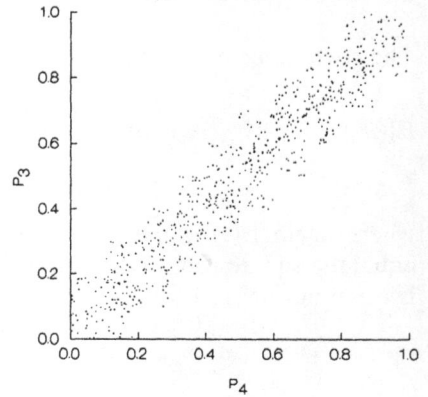

(b)

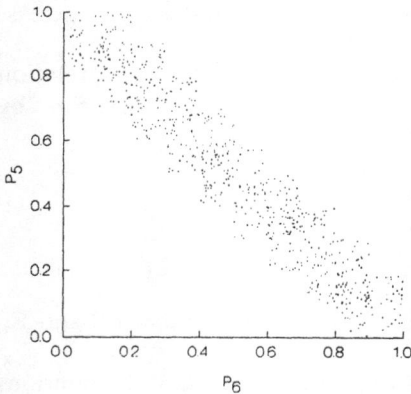

(c)

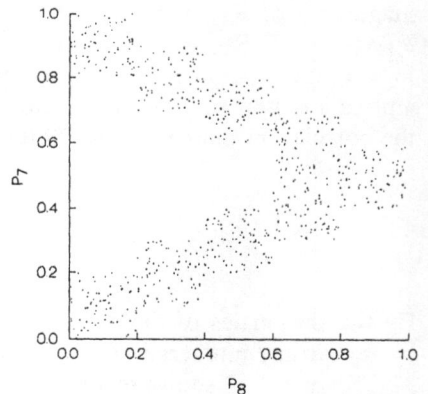

(d)

Figure 2.13 Scatter plots for jointly distributed random parameters where the parameters are (a) uncorrelated, $R_{12} = 0.034$; (b) positively correlated, $R_{34} = 0.953$; (c) negatively correlated, $R_{56} = -0.951$; and (d) uncorrelated, $R_{78} = 0.013$.

rameter value. Mathematically this means $\mathcal{P}_1$ and $\mathcal{P}_2$ are independent if $f(P_1, P_2) = f(P_1)f(P_2)$. In this case, $f(P_1)$ and $f(P_2)$ are the marginal densities of $\mathcal{P}_1$ and $\mathcal{P}_2$, respectively.

Figure 2.14 shows scatter plots for two two-dimensional data sets. This data is taken from Purviance [4]. The data sets have the same marginal densities and covariance, but those in (a) are uncorrelated, while those in (b) are independent. Viewing these data sets, you can see that they are statistically different, but the difference is not measured either in the marginal densities or in the correlations.

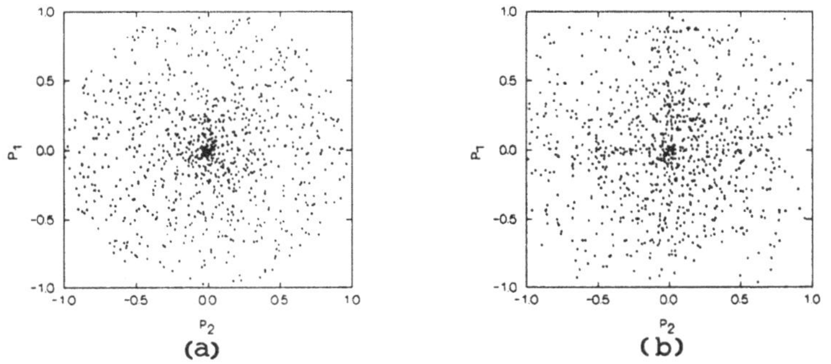

Figure 2.14 Scatter plots for two data sets with the same marginal densities and the same covariance. Both are uncorrelated; however, they are not statistically identical. Data set (a) is restricted to a unit circle centered at 0,0; data set (b) does not have this restriction.

Beware of using marginal densities and correlations to determine if two data sets are statistically identical.

From the scatter plots in Figure 2.14 we can see that "independent parameters" does not mean the same as "uncorrelated parameters." By definition, independent parameters are also uncorrelated, but uncorrelated random parameters are not necessarily independent. However, for the special case of jointly Gaussian random parameters, the two are equivalent.

2.4.10 Higher-Order Statistics

In equation (2.10), we defined the covariance between two random parameters as $E\{(\mathscr{P}_1 - P_{10})(\mathscr{P}_2 - P_{20})\}$. This covariance is a special case of a general covariance, $COV_{m,n}(\mathscr{P}_1, \mathscr{P}_2)$, of the form

$$COV_{m,n}(\mathscr{P}_1, \mathscr{P}_2) = E\{(\mathscr{P}_1 - P_{10})^m (\mathscr{P}_2 - P_{20})^n\}$$

We call these statistics higher-order statistics. The differences between the two data sets shown in Figure 2.14 can be accounted for by the differences in their higher-order statistics.

There are some very useful ways to determine parameter statistics from samples of a random parameter. Samples of a random parameter are generated by performing repeatedly the underlying experiment that generates the random parameter, and recording the parameter value for each experiment. An important assumption made here about sampling is *sample independence*. In essence, sample independence is achieved when knowledge of past samples gives no new knowledge about any future samples, aside from the underlying density information [2], no matter how many past samples are considered. The manufacturing model described assures sample independence, assuming the sampling from the bins is truly random, and the bins contain a large number of components. When we sample a random parameter, the parameter sample is generated according to its PDF. There is always a statistical consistency given to the parameter samples, as described by the parameter PDF. One of the ways to determine the statistical properties of the underlying PDF is to estimate the parameter statistics, like mean and variance, from the parameter samples.

Assume we have made N samples of the random parameter $\mathcal{P}$. Call these P_i, i = 1 to N. This collection of N samples forms a parameter sample database. Now let $\hat{P}_0$ be the unbiased estimate of the mean value P_0 (an unbiased estimate converges to the real estimate in the limit as N, the number of samples, goes to infinity). Then,

$$\hat{P}_0 = \frac{1}{N}\sum_{i=1}^{N} P_i$$

A table of unbiased estimates for some of the moments of a random parameter are given in Table 2.1.

Table 2.1
The First Four Moments of the Random Parameter $\mathcal{P}$, and Their Unbiased Estimates

$E(\mathcal{P}) = P_0 =$ mean	$$\hat{P}_0 = \frac{1}{N}\sum_{i=1}^{N} P_i$$
$E\{(\mathcal{P} - P_0)^2\} = \sigma^2 =$ variance	$$\hat{\sigma}^2 = \frac{1}{N-1}\sum_{i=1}^{N}(P_i - P_0)^2$$
$E\{(\mathcal{P} - P_0)^3\} = \mu_3 =$ skewness	$$\hat{\mu}_3 = \frac{N}{(N-1)(N-2)}\sum_{i=1}^{N}(P_i - P_0)^3$$
$E\{(\mathcal{P} - P_0)^4\} = \mu_4 =$ kurtosis	$$\hat{\mu}_4 = \frac{N^2 - 2N + 3}{(N-1)(N-2)(N-3)}\sum_{i=1}^{N}(P_i - P_0)^4$$

2.5 GEOMETRIC APPROACH TO YIELD CALCULATION

Up to now in this chapter we have been laying the foundation necessary to describe yield geometrically in parameter space. The key to this description is the parameter acceptability region, $\mathcal{P}_a$. Embedded in the parameter acceptability region definition are the measurement specifications, S, and the circuit performance function, G(P). In this context, a unit passes performance testing when the manufacturer of the unit chooses component parameter values from $\mathcal{P}_a$. Thus the success of unit manufacture will measure our ability to appropriately and accurately model the linkage between the design and manufacturing processes.

2.5.1 The General Geometric Approach

Recall that we have three equivalent definitions for yield:

1. The ratio of the number of manufactured units which pass performance testing to the total number of units manufactured, in the limit as the number of units goes to infinity.
2. The probability that a manufactured unit will pass its performance tests.
3. The probability that the parameter choice made during manufacture lies within the parameter acceptability region for that design.

This third definition gives rise to the geometric approach to yield calculation.

A way of calculating the probability of choosing a parameter from the acceptability region is to determine the area, or volume, under the joint parameter PDF (i.e., integrate) over the parameter acceptability region. Numerically, this geometric definition makes sense because we have defined the total area under f(P) to be always 1.0. If f(P) lies totally within the acceptability region, all parameters are chosen from $\mathcal{P}_a$, all of the manufactured units meet specification, and the yield is 100%. If f(P) lies totally outside of the acceptability region, then no units pass performance testing, and the yield is 0.0%. For the in-between cases, the area must somehow be determined. Figure 2.15 shows the graphical representation of a two-dimensional problem, with $\mathcal{P}_a$, $f(P_1,P_2)$, and P_{10} and P_{20}, the parameter nominal values in parameter space.

Three ways to increase the yield of a unit were discussed in Section 1.6.2. The first is for the manufacturing engineer to control tightly the variation of the manufacturing parameters. This type of control is mathematically equivalent to decreasing the extent of f(P). This will often increase the percentage of f(P) within $\mathcal{P}_a$. However, in practice, decreasing the extent of f(P) can be a very costly process, because tight control of the manufacturing process may not be easy. For example, buying 1% resistors usually costs more than buying 10% resistors.

In a second approach, the designer can use a unit design structure whose performance is insensitive to the uncontrollable manufacturing parameters. The proper choice of design structure is equivalent to increasing the size of $\mathcal{P}_a$. This will always increase the percentage of f(P) within $\mathcal{P}_a$, and hence will always increase the yield. However, we assume that the designer has already chosen a "good" structure for the design.

With the third approach, the designer can choose nominal parameter values, P_{10} and P_{20}, so that the parameter joint density better covers $\mathcal{P}_a$. This is equivalent

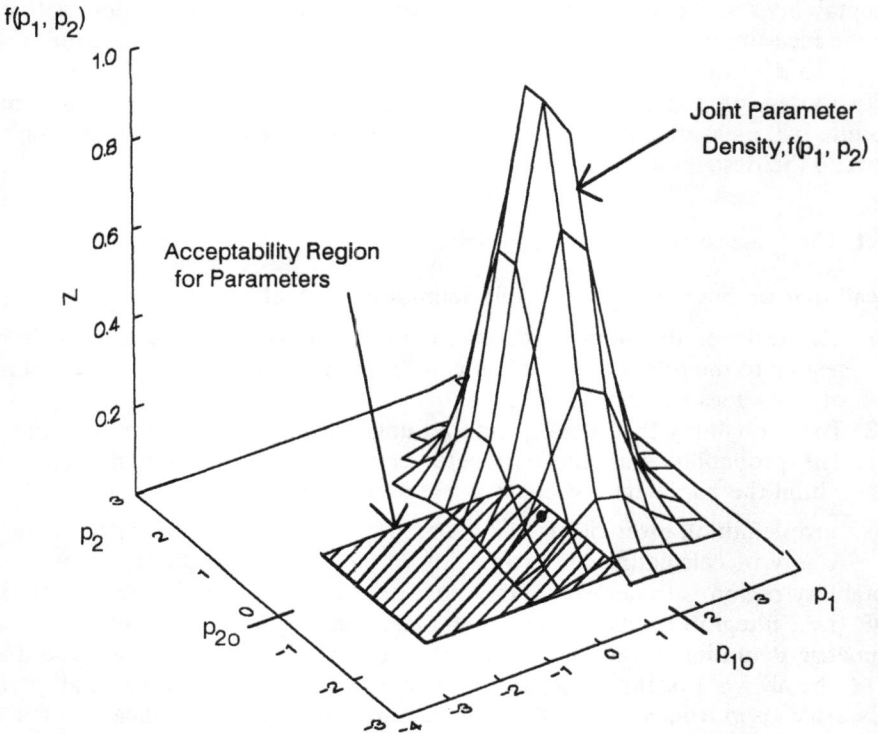

Figure 2.15 Relationship between the position of a joint parameter density function, and the position of the acceptability region in parameter space.

to moving $f(P_1, P_2)$ around in parameter space so that the area under $f(P_1, P_2)$ is maximized over the acceptability region. Although all three of these strategies should be used to obtain high yield, we will concentrate on the third approach, as it is the essence of statistical design.

Looking at the simple two-dimensional example given in Figure 2.15, it may seem that statistical circuit design is simple—just place the nominal unit parameters in the "center" of $\mathcal{P}_a$ to obtain the highest yield. Although this is the right way to think, the center of $\mathcal{P}_a$ may not always give the highest yield design because it may not take into account instances in which the parameter distributions are skewed or multimodal. Also, it is generally the case that $\mathcal{P}_a$ is not a regular shape, and therefore, identifying the center may be impossible. Furthermore, $\mathcal{P}_a$, and hence its center (if defined), is not directly calculable because the performance function, $G(P)$, is not invertible. The solution to the problem of how to place the nominal design for highest yield is one of fitting, in an optimal way, a given parameter joint density to a given acceptability region, under the realization that the acceptability region is not directly available by calculation. We hope by this discussion to make clear that the geometric interpretation is very useful. We will use it extensively as we study the different approaches to statistical unit design.

2.5.2 Yield With Uniform Independent Parameters

When the parameters of a unit are each uniformly distributed and jointly independent, the geometric interpretation for yield is particularly simple. For this case, each parameter has a fixed extent of variation, and any particular parameter value within this variation interval is equally likely to be chosen. Hence, the tolerance region in $\mathcal{P}$ is well-defined and all of the parameters within the tolerance region are equally likely to occur. Consequently, f(P) and $\mathcal{T}$ cover the same extent, and f(P) is constant over $\mathcal{T}$.

For this case, yield is simply the area (or volume for higher dimensional problems) of the overlap of the acceptability and tolerance regions, divided by the area (or volume) of the tolerance region. This is given as

$$\text{Yield} = \frac{\text{area}(\mathcal{T} \cap \mathcal{P}_a)}{\text{area}(\mathcal{T})} \qquad (2.12)$$

where area (.) denotes the area, or volume, of the enclosed region. Figure 2.16 shows a two-dimensional example of yield calculation when parameters are jointly

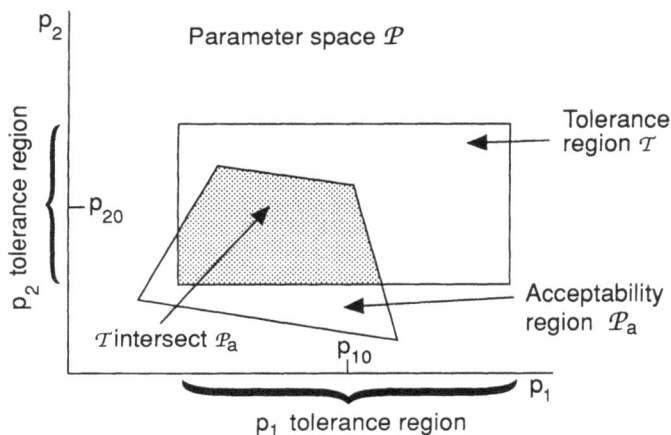

Figure 2.16 Two-dimensional plot of yield calculation for jointly uniform, independent parameters. Yield is the area of the intersection of $\mathcal{T}$ and $\mathcal{P}_a$, divided by the area of $\mathcal{T}$.

uniform and independent. From this figure, it is very easy to visualize the yield-optimization problem. The nominal values of the parameters, P_{10} and P_{20}, are chosen by the designer to maximize the overlap of T with the acceptability region. Although in practice parameter values are not uniform and independently distributed, this simple geometric concept of yield as the area (or volume) of the overlap of $\mathcal{P}_a$ and T is useful for visualizing more complex yield-calculation and optimization problems. The two-dimensional example developed here can form a very strong foundation for understanding statistical design and optimization. It should be noted that when f(P) is not composed of uniform and independent densities, these results do not exactly apply.

All of these ideas can be extended to multidimensional problems. In these cases the tolerance and acceptability regions are volumes in n-dimensional parameter space. For multidimensional, jointly uniform, independent parameters, yield is the volume of the intersection of $\mathcal{P}_a$ and T divided by the volume of T.

2.5.3 Example—A Voltage Divider

For the voltage divider used as an example in the previous sections, the yield can be geometrically calculated. The parameter space for this circuit and $\mathcal{P}_a$ from Figure 2.6 are repeated in Figure 2.17. Given the resistor data where each is $35\Omega \pm 5\Omega$,

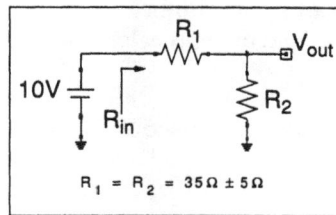

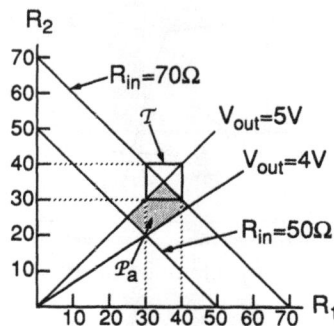

Figure 2.17 Geometric yield calculation of approximately 25% for a voltage-divider circuit with uniform, independently distributed parameters.

uniform and independently distributed, the tolerance region can be plotted. Looking at the overlap of the tolerance region and the parameter acceptability region, we see that the yield is approximately 25%.

2.6 MATHEMATICAL APPROACH TO YIELD CALCULATION AND YIELD AS A MULTIDIMENSIONAL INTEGRAL

Although the geometric approach to yield provides a strong intuitive understanding, it does not always lend itself to rigorous mathematical treatment. For this reason, another formulation of yield is useful for calculations and applications which require additional mathematical rigor.

From the geometric discussion of yield, we can identify the information necessary to determine yield as f(P), the joint parameter statistics, and $\mathcal{P}_a$, the acceptability region. The parameter statistics contain the manufacturing information, and the acceptability region contains the testing specifications and unit-performance information. Using a multidimensional integral, yield is related to these quantities as

$$
\begin{aligned}
\text{Yield} &= \int_{\mathcal{P}_a} f_{\mathcal{P}}(P)dp \\
&= \iint_{\mathcal{P}_a} \cdots \int f_{\mathcal{P}}(p_1, p_2, \ldots p_n)dp_1 dp_2 \ldots dp_n
\end{aligned}
\tag{2.13}
$$

It is important to note that the integral is only applied over the parameter acceptability region. The calculation of yield requires the knowledge of f(P) and the shape of $\mathcal{P}_a$ and an ability to integrate over these n-dimensional regions. In practice, neither is known exactly, so this formulation for the yield is generally not practical to use.

A slightly more useful expression for yield involves the testing function, accept(P), given by

$$
\text{accept}(P_i) = \begin{cases} 1 \text{ if } P_i \text{ is a member of } \mathcal{P}_a \\ 0 \text{ if } P_i \text{ is not a member of } \mathcal{P}_a \end{cases}
\tag{2.14}
$$

where $P_i = (p_1, p_2, \ldots p_n)$ is a point in parameter space. Yield can then be expressed as an expectation with respect to accept(P) as

$$
\begin{aligned}
\text{Yield} &= E\{\text{accept}(P)\} = \int_{-\infty}^{\infty} \text{accept}(P)f(P)dP \\
&= \int_{-\infty}^{\infty} \int_{-\infty}^{\infty} \cdots \int_{-\infty}^{\infty} \text{accept}(p_1, p_2, \ldots, p_n)f_{\mathcal{P}}(p_1, p_2, \ldots, p_n)dp_1 dp_2 \\
&\quad \ldots dp_n
\end{aligned}
\tag{2.15}
$$

This formulation in itself is difficult to use because of two unfortunate practical facts:

1. Generally, an analytic form for the parameter density is not known.
2. An analytical form for accept(P) is usually not available.

However, $\text{accept}(P_i)$ can be evaluated for any P_i by determining if $G(P_i)$ meets the unit test specifications; that is, $\text{accept}(P_i) = 1$ if and only if $G(P_i)$ is within the measurement specifications for the unit. This is encouraging in some sense, because our ability to calculate $\text{accept}(P_i)$ for any given P_i rests on our ability to determine unit performance for the parameter set P_i, and to compare the calculated performance to the unit test specifications. Fortunately, there are many good analytic simulators which can accurately determine unit performance as a function of component parameters. In the next chapter, we use these simulators as a starting point for Monte Carlo integration of multidimensional integrals.

2.7 CONCLUSION

Our primary focus in this chapter has been to identify and discuss methods for gaining insight into how parametric yield is defined and calculated. We have identified two approaches for the characterization of yield. The geometric formulation of yield appeals to our intuition by providing a straightforward, graphical environment where we can actually see yield. The mathematical approach provides the analytic underpinnings necessary for the implementation of yield calculation and optimization algorithms. Each method for describing yield lends value to the other—they are both important to the practitioner of robust design techniques. In addition, both approaches require a short review of random variables, leading up to the realization that the parameter joint probability density function $f_{\mathcal{P}}(P)$ is crucial for complete characterization of the manufacturing process.

In Chapter 3, we extend the ideas just presented and learn the particulars of the available yield-calculation algorithms and codes.

2.8 IMPORTANT IDEAS FROM CHAPTER 2

Section 2.1

- Yield is conceptually defined as the ratio of the number of manufactured units that pass performance testing to the total number of manufactured units, in the limit as the number of manufactured units approaches infinity.

Section 2.2

- There are two different approaches to the description of yield: geometry and mathematics. The geometric approach is visual and intuitive. The mathematical description is necessary for the analytic formulation of yield-estimation and optimization algorithms.

Section 2.3

- Yield can easily be viewed in terms of two spaces, parameter space $\mathcal{P}$, and performance space $\mathcal{M}$. Any realization of a unit can be represented as a point in parameter space, $\mathcal{P}$, and the unit's performance is a point in performance space $\mathcal{M}$.
- Associated with each measurement in $\mathcal{M}$ is a design specification S, which, in turn is defined by an inequality or inequalities (i.e., 25dB < gain < 28dB).
- The acceptable performance region $\mathcal{M}_a$ is realized by the collection of all possible performance points in M that fall within specification.
- The performance function G(P) maps points in parameter space to performance space. G(P) is typically very nonlinear and usually maps 1:1 but not "onto."
- The acceptable region in parameter space $\mathcal{P}_a$ is defined as the collection of all points in parameter space that, when mapped to performance space fall within $\mathcal{M}_a$, the acceptable performance region.
- The tolerance region $\mathcal{T}$ is defined as the region in P over which realizable values of parameters are possible.

Section 2.4

- To effectively model the manufacturing environment, the joint parameter statistical variations must be accurately captured.
- The mean or average value is best envisioned as a center-of-gravity locator. In this way, it is easy to explain why a bimodal distribution can involve an average value which seldom or even never occurs.
- For several random variables, the linear association between any pair is conveniently represented in matrix form by the correlation matrix $\mathbf{C}$.
- Mean values, standard deviations and pair-wise linear correlation coefficients are necessary, but not sufficient, to model typical joint parameter probability functions.
- The practical realization of the integral moment equations is accomplished through the unbiased estimate forms for the corresponding moment.

Section 2.5

- From a geometric viewpoint, the yield-optimization problem is one of fitting, in an optimal way, a given parameter joint density to a given acceptability region, under the realization that the acceptability region is not directly available by calculation.
- The parameter joint density function is rarely available in analytic form, making the yield optimization problem even more illusive.

Section 2.6

* Practical evaluation of the multidimensional yield integral is accomplished via Monte Carlo integration techniques.

REFERENCES

[1] J. Turino, "Concurrent Engineering, Making It work calls for Input From Everyone," *IEEE Spectrum,* Vol. 28, No. 7, July 1991, pp 30-32.

[2] L. Devroye, *Non-Uniform Random Variate Generation,* New York, NY: Springer-Verlag, 1986.

[3] M. D. Meehan and L. Campbell, "Statistical Techniques for Objective Characterization of Microwave Device Statistical Data," *Proc. of the 1991 MTT-S Int. Microwave Symposium,* Boston, MA, June 1991, pp 1209-1212.

[4] J. Purviance, "Design Centering for Yield Improvement of an HEMT LNA," *Proceedings of EEsof Users' Group Meeting,* June 14, 1989.

Chapter 3
Calculating Yield

"Work smarter, not harder."

3.1 INTRODUCTION

The goal of design for manufacture is to account for manufacturing uncertainties throughout the design process. Often this involves maximizing the manufacturing yield. Because it is almost always necessary to use the computer to perform this maximization yield, or at least some function proportional to it, must be mathematically calculated. In this chapter we will sort through and classify many methods that are available to calculate yield. Each yield-calculation technique has strengths and weaknesses, and we will present the benefits and drawbacks for several of the most popular methods.

You may be of the opinion that getting involved with yield calculation entails too much detail, if the goal is simply to be a user of commercial statistical-design software. However, this is not true. The subtleties of yield calculation directly affect both accuracy and execution speed of the software. Today, in order to run any commercial software, the user must enter certain program variables. These variables include:

1. How many points to use in a Monte Carlo yield calculation;
2. How many auxiliary circuits to use in a yield optimization;
3. Whether to use a variance reduction technique like the shadow model.

Therefore the user must get involved with some of the calculation aspects of the science. At this time, yield-optimization algorithms, although very good, are

---

not structured to be used in a turn-key manner; user interaction and understanding is required for best results.

This chapter will also be useful if you plan to write your own, or modify existing statistical design software. Although an introduction to calculating yield is given, we cannot present enough detail here to allow you to write efficient yield-calculation algorithms. But you will be exposed to pertinent literature describing the latest state-of-the-art yield-calculation techniques.

So what do we have to know and do to calculate yield? As explained in Chapter 2, yield is mathematically equal to an n-dimensional definite integral where n is the number of parameters used to characterize a unit's performance:

$$\text{Yield} = \int_{\mathcal{P}_a} f_{\mathcal{P}}(P)dP \tag{3.1}$$

This formulation requires knowledge of both the parameter statistics, $f_{\mathcal{P}}(P)$, and the acceptability region, $\mathcal{P}_a$. However, determining $\mathcal{P}_a$ exactly is impossible, except in very simple cases, because the parameter acceptability region generally does not have a concise, closed-form mathematical description. Determining the acceptability region requires knowledge of the acceptance function, accept(P). Calculation of yield using accept(P) leads to an alternative integral expression for yield,

$$\text{Yield} = \int_{-\infty}^{\infty} \text{accept}(P)f_{\mathcal{P}}(P)dP \tag{3.2}$$

Accept(P) involves two elements, the performance function G(P), and the unit specifications, S. Once again though, an analytic form for G(P) is not known. However there are software simulators that can be used to evaluate $G(P_i)$ in a point-wise fashion, for any value of P_i. But still, notice that a closed-form expression for accept(P_i) is nonexistent because the unit performance solver typically relies on matrix manipulations to solve systems of equations, and perhaps even iterative techniques.

As we shall see, each of these two expressions for yield, (3.1) and (3.2), motivate different approaches for calculating yield, many of which will be explored in Section 3.6.

While the main purpose of this chapter is to present various methods for yield calculation, we first provide a preliminary study of mathematical Monte Carlo methods (Section 3.2). The fact that most methods use Monte Carlo calculation in some capacity provides the motivation behind this presentation. In Section 3.3, the Monte Carlo concepts are extended within the framework of yield calculation. Next, the geometric aspects of yield calculation (see Section 2.5) are briefly revisited (Section 3.4). Section 3.5 develops a novel classification scheme which will be used to compare the methods presented in Section 3.6.

3.2 MONTE CARLO INTEGRATION

3.2.1 Fundamental Theorem of Monte Carlo

There is a mature area of numerical mathematics which deals with the evaluation of multidimensional definite integrals of the form

$$G = \int_{-\infty}^{\infty} G(P)f_{\mathscr{P}}(P)dP \tag{3.3}$$

where

$$f_{\mathscr{P}}(P) \geq 0, \quad \text{and} \quad \int_{-\infty}^{\infty} f_{\mathscr{P}}(P)dP = 1 \tag{3.4}$$

The fundamental theorem of Monte Carlo guarantees that the following game of chance may be used to make numerical estimates of G [1]:

Choose a set of vectors P_1, P_2,...,P_n from $f_{\mathscr{P}}(P)$ such that the P_i are samples of the joint PDF $f_{\mathscr{P}}(P)$. The function $f_{\mathscr{P}}(P)$ can be thought of as a density because it must be nonnegative and have a volume of one. Then, if we evaluate $G(P)$ at each of the sample vector points, P_i, we can form the arithmetic mean as:

$$G_N = \frac{1}{N}\sum_{i=1}^{N} G(P_i) \tag{3.5}$$

Then the Fundamental Theorem of Monte Carlo states that

$$\lim_{N \to \infty} G_N = G \tag{3.6}$$

if G exists.

In general if $f_{\mathscr{P}}(P)$ is not a density, then it can be multiplied by a normalizing function, $N(P)$, such that the product $f_{\mathscr{P}}(P)N(P)$ has the properties of a density. The function $G(P)$ can be divided by $N(P)$, and hence, the value of G is not changed. (A very useful variation on this idea is *importance sampling* which will be examined later in this chapter.)

Notice that the result stated in the fundamental theorem is completely independent of the dimension of P and the form of $G(P)$ or $f_{\mathscr{P}}(P)$. This technique offers us a powerful tool for evaluating complicated multidimensional definite integrals. Furthermore, $G(P)$ need not be known analytically, it only needs to be evaluated at the sample vector points, P_i.

For a finite number of sample points N, G_N is an estimate of G. Therefore, we can write

$$G_N = G + \text{error}$$

If G exists, the error is a random variable with a zero mean, and the error magnitude is bounded with high confidence by

$$|\text{error}| \text{ is bounded by } \frac{\sigma}{\sqrt{N}}$$

for large N, and

$$\sigma^2 = \int_{-\infty}^{\infty} G(P)^2 f_{\mathcal{P}}(P)dP - G^2 \qquad (3.7)$$

Finally, a useful relationship for N in terms of the error and the variance is

$$N = \frac{\sigma^2}{|\text{error}|^2} \qquad (3.8)$$

This relationship will be employed later in this chapter for developing confidence intervals for the estimate of G. The fact that error is a zero-mean random variable means that as N becomes large, error goes to zero. This desirable condition makes G_N an unbiased estimate of G.

3.2.2 Ratio of Volumes Interpretation

For the case where G(P) is a binary function assuming the values 0 or 1, there is an excellent method that will help visualize this concept by giving Monte Carlo integration a geometric interpretation. Consider the problem of determining the ratio of two volumes, where one volume is larger and totally contains the other volume. An approximation to the ratio of the two volumes can be made by choosing points within the biggest volume at random (i.e., uniform sampling). An estimate of this volume ratio is

$$\text{volume ratio estimate} = \frac{\text{number of points within small volume}}{\text{total number of points chosen}}$$

This equation also results in an unbiased estimate.

A one-dimensional illustration of this ratio is to consider two intervals on the real line.

Interval-a: $1 < P < 2$
Interval-b: $0 < P < 3$

If we wish to determine the length ratio,

length of Interval-a
length of Interval-b

we need only to uniformly choose many points on Interval-b and take the ratio of the number of points which fall on Interval-a to the total number of points chosen. (You can easily do this on paper and it works.) Approximately one third of the points chosen will lie within Interval-a. If you expand this concept to two and three dimensions, you can determine the ratio of areas and volumes respectively. As can be seen, the dimensionality of the problem does not affect the accuracy of the estimate.

Figure 3.1 illustrates the choosing of a point at random within a one-, two-, and three-dimensional volume. Let the total volume be somehow divided into two equal volumes. Then, the probability of a uniformly random point lying in one of the two equal volumes is 0.5. This result is independent of n (the dimension of the volume), independent of the shape of the volume, and independent of the geometry of the division of the volume into two equal parts.

Some numerical procedures become significantly more complex as they are applied to higher dimension problems. This complexity is called the "curse of dimensionality" [2]. Monte Carlo estimation techniques do not suffer from the curse of dimensionality.

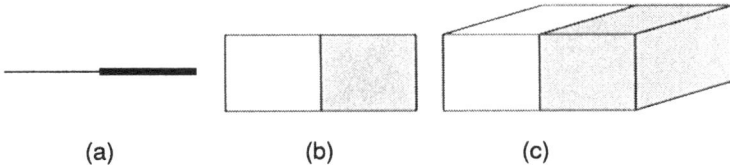

(a)   (b)   (c)

Figure 3.1 Three figures: (a) a line (one dimension), (b) a square (two dimensions), and (c) a rectangular solid (three dimensions), each divided into equal "volumes."

3.2.3 The Definite Integral of a Binary Function

One can apply the Fundamental Theorem to the problem of estimating the definite integral of $G(P)$ over the region A, where $G(P)$ is a binary function—either 0 or 1. The integral of $G(P)$ is

$$\int_A G(P)dP = G \qquad (3.9)$$

This gives a definite integral over just the volume A instead of over the entire space as stated in the fundamental theorem of Monte Carlo. In order to use the fundamental theorem, one first needs to change the form of the integral slightly. Let us write G as the definite integral with infinite limits:

$$G = \int_{-\infty}^{\infty} \frac{G(P)}{\frac{1}{K}} \frac{1}{K} f(P)dP \qquad (3.10)$$

where

$$f(P) = \begin{cases} 1 \text{ if P is a member of A} \\ 0 \text{ otherwise.....} \end{cases}$$

and K is some constant such that

$$\int_A \frac{1}{K}dP = 1$$

that is, K is the volume of A.

Interpreting $1/K\, f(P)$ as a uniform density defined over the volume A, the Fundamental Theorem of Monte Carlo can now be applied and it states that

$$G_N = \frac{1}{M}\sum_{i=1}^{M} \frac{G(P_i)}{\frac{1}{K}} \qquad (3.11)$$

where G_N is an unbiased estimate of G, and the P_i are chosen uniformly from A.

The volume ratio is given by:

$$\frac{\text{volume of } G(P) \text{ within A}}{\text{volume of A}} = \frac{\int_A G(P)dp}{\int_A dP} = \frac{G}{K}$$

And an unbiased estimate of this volume ratio is given by

$$\frac{G_N}{K} = \frac{1}{M}\sum_{i=1}^{M} G(P_i) \tag{3.12}$$
$$= \frac{\text{number of points where } G(P_i) = 1}{\text{total number of points chosen}}$$

where the P_i points are uniformly chosen from region A, the larger region. Using the Fundamental Theorem to evaluate the ratio of the two volumes in this manner is similar to solving the yield integral because yield can be defined as a weighted volume ratio. Notice that this technique works for any binary function $G(P_i)$. The function does not need to be connected or convex (see Section 3.5) or have any special properties. Often A can be chosen so that its volume, K, is known or easy to calculate. (The only restriction on A is that it must enclose $G(P)$). This general property of Monte Carlo integration will be used to calculate the yield of circuits and systems. (See Exercise 3.1.)

3.3 MONTE CARLO APPROACH TO YIELD CALCULATION

The yield integral, for parameters distributed by the parameter joint PDF $f_{\mathcal{P}}(P)$, is given by (3.2).

$$Y = \int_{-\infty}^{\infty} \text{accept}(P)f_{\mathcal{P}}(P)dP$$

Applying the Fundamental Theorem of Monte Carlo to this integral results in an unbiased estimator of the yield integral, Y, given by

$$\hat{Y} = \frac{1}{M}\sum_{i=1}^{M} \text{accept}(P_i) = \frac{k}{M} \tag{3.13}$$

where P_i is a sample of the parameter values taken from the parameter PDF $f_{\mathcal{P}}(P)$, k is the number of accepted units, and M is the total number of units tested. The variance of $\hat{Y}$ is proportional to $1/M$. This unbiased estimator forms the basis for Monte Carlo yield estimation. If you use this formulation for calculating yield, you can simulate a large number of units in which the parameters for each unit are samples from the parameter PDF. Then the performance of each unit can be tested against its performance specifications. The Fundamental Theorem states that the fraction of units which meet specification is the yield estimate.

Monte Carlo yield-estimation techniques are very useful because

- They do not require an analytical description of either $\mathcal{P}_a$ or $\mathcal{T}$.

- The variance of the yield estimate is independent of the number of parameters associated with the circuit [3].

The yield estimate given by the Fundamental Theorem is also very intuitive since it simulates the manufacturing and test models that are being used. By choosing the parameter vectors according to the modeled joint density function we imitate the values assigned to parameters during manufacture. Testing manufactured units against the performance specifications to determine if each passes or fails is equivalent to evaluating accept(P_i) for each Monte Carlo trial. Taking the sum of all trials having accept(P_i) = 1, and dividing by the number of units simulated equates to determining the fraction of units which will pass performance testing. The general Monte Carlo yield-estimation algorithm is illustrated in Figure 3.2.

We would like to stress that this intuitive interpretation of yield is very useful for understanding and using Monte Carlo yield estimation. However, Monte Carlo

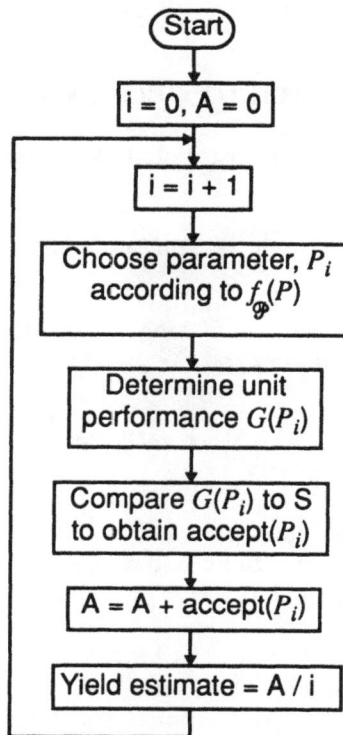

Figure 3.2 Monte Carlo yield-estimation algorithm.

yield estimation is much better founded than simply by our intuition. These ideas come from an important branch of numerical mathematics, where systematic thought and rigorous analysis have been applied.

Monte Carlo is an established technique for estimating the value of multidimensional integrals, and has a wide range of applications in applied mathematics. It is a powerful tool in that the estimate it provides requires no assumptions on the forms of $f_{\mathcal{P}}(P)$ or $\mathcal{P}_a$. Its only drawback is that it requires a large number of trials to obtain an accurate estimate. Thus, the computational burden can be significant. To reduce this burden, several important techniques have been developed to reduce the variance in Monte Carlo estimates [3, 4]. We will discuss some of these in the following sections.

3.3.1 Confidence Intervals

Because $\hat{Y}$ is a statistical estimate of yield, it is itself a random variable. Fortunately the characteristics of $\hat{Y}$ are predictable. As an example, suppose we were to perform a series of K experiments to compute $\hat{Y}$, wherein each estimate $\hat{Y}$ contained the same number of trials, M. Using a computer with a different random-number-generator seed for each experiment would result in varying estimates for $\hat{Y}_1$ to $\hat{Y}_K$. In fact, if each estimate consists of greater than 40 trials [7], then in the limit as K $\rightarrow$ ∞, the distribution of the difference between the actual yield, Y, and the yield estimate, $\hat{Y}$, can be regarded as a zero mean Gaussian random variable. A deviation range taken symmetrically about zero (the actual yield) is termed the *confidence interval*. The *confidence level* is defined as the probability that yield estimate $\hat{Y}_i$, determined from M trials will fall within a given confidence interval. For example, given a certain confidence level, say 95%, we can determine an interval where, on the average, 95 out of 100 of the $\hat{Y}_i$s will fall. Note that this interval is both a function of Y, the actual yield, and M, the number of trials. The extent of the confidence interval for $\hat{Y}$ can be computed directly using

$$|\hat{Y} - Y| \text{ is bounded in confidence by: } Z_c\sqrt{\frac{Y(1 - Y)}{M}} \qquad (3.14)$$

where Y is the actual yield ($0 <$ Yield < 1), $\hat{Y}$ is the Monte Carlo yield estimate, M is the number of Monte Carlo samples, and Z_c is the confidence level variable. This confidence interval calculation assumes that M is large enough that $Y > 0.5$ and $M(1 - Y) > 5$, or that $Y < 0.5$ and $MY > 5$ [4].

The radical term in (3.14) modifies the confidence interval for a standard normal random variable—Z_c. Thus Z_c is 2.58 for 99% confidence, 1.96 for 95% confidence, 1.28 for 80%, and 0.675 for 50% confidence. Table 3.1 presents typical values of $|\hat{Y} - Y|$ for different values of Z_c, Y, and M. In practice $\hat{Y}$ is substituted for Y in the third column of Table 3.1 [7].

Table 3.1
Error in the Yield Estimate for Various Combinations of Z_c, Y, M and Confidence Level

Confidence Level	Z_c	Y	M	solve for $|\hat{Y} - Y|$	Error
99%	2.58	0.8	1000	0.033	3.3%
80%	1.28	0.6	500	0.028	2.8%
95%	1.96	0.7	2000	0.020	2.0%
80%	1.28	0.7	2000	0.013	1.3%
95%	1.96	0.7	200	0.064	6.4%
95%	1.96	0.7	20	0.20	20.0%

This table shows that for a 95% confidence in the estimate, yield = 0.7, and 2000 Monte Carlo trials, the estimated yield will have an absolute error of less than 2.0%. That is to say on the average, 95 out of 100 yield calculations (M = 2000) will be between 0.72 and 0.68. The same situation with 200 Monte Carlo trials gives an absolute error less than 6.4%, and with 20 Monte Carlo trials gives an absolute error of less than 20%. This illustrates that the confidence in the yield calculation degrades as the number of Monte Carlo trials becomes smaller. In practice, 1000 or more trials are usually necessary if we are to use the estimate as a reliable interpretation of a circuit's yield.

3.3.2 Variance Reduction

Many researchers have addressed the computational burden associated with Monte Carlo integration. Any technique that accelerates convergence of a Monte Carlo computation is classified as a *variance-reduction* technique. This is because the variance of the yield estimate for a given number of Monte Carlo calculations is reduced. Two representative variance-reduction methods are now presented.

Control-Variate Variance Reduction

The control-variate (CV) technique is a general statistical method for variance reduction [4] that helps to reduce the computational burden of Monte Carlo. Application of the CV technique to electrical circuits began with Neill [5], and was further refined by Hocevar, Lightner, and Trick [3], and Soin and Rankin [6]. Successful application of this technique hinges on the identification and synthesis of an "auxiliary" simulation model, referred to as the *shadow model*.

Shadow Model

The shadow model is simply a surrogate unit-performance simulator having the desirable characteristic that it is computationally more economical than the original unit-performance simulator. The shadow model contains parameters that correspond to those of the original unit. Changes in these parameters produce nearly equivalent changes in the response of both the original unit and shadow model. The vehicle for this variance reduction can be easily identified with the following mathematics.

Start by decomposing the yield definition into two components,

$$Y(P_0) = \check{Y}(P_0) + \Delta Y(P_0)$$

where $\check{Y}(P_0)$ is the yield of the shadow model and $\Delta Y(P_0)$ is the yield error represented by the difference between the actual yield and the shadow model yield. An estimator of the yield is then given by

$$\hat{Y}(P_0) = \hat{\check{Y}}(P_0) + \Delta Y(P_0)$$

Due to the computational savings afforded by the use of the shadow model, $\hat{\check{Y}}(P_0)$ is inexpensive to compute and can be calculated with little statistical error. Consequently the error in the control variate estimator lies almost entirely in the term $\Delta Y(P_0)$. For the same computational effort, Hocevar et al. [3] have shown that when the correlation coefficient between the shadow model and the simulated performance of the unit is high (i.e., the shadow model is a close predictor of the unit's performance), the variance of the control variate yield estimator is smaller than the variance of the conventional Monte Carlo yield estimate.

However, you should note that when using the shadow model, there is a trade-off between model accuracy and computational efficiency. A large amount of CPU time may be initially required to obtain a shadow model that is accurate enough to give variance reduction. To obtain variance reduction the model must have a correlation coefficient of 0.5 or better [6]. Hocevar et al. [3] derived an efficiency metric which accounts for this accuracy-computation trade-off:

$$\eta = \frac{\sigma_S^2 T_S}{\sigma_{CV}^2 T_{CV}}$$

where the subscripts (S and CV) correspond to standard and CV techniques respectively, for the yield-estimate variance, σ^2, obtained in T seconds.

The main challenge with the shadow-model method is in identifying a shadow model which follows the original circuit "closely", yet has negligible computational effort as compared to the original unit simulator. Researchers have developed several candidates for the shadow model and examples of them are given in recent literature [7, 6]. In addition, Hocevar et al. [3] used a quadratic approximation to the circuit response surfaces and reported correlation coefficients and efficiency ranges from 0.88 to 0.99 and 7.3 to 23.1, respectively. Presently, commercial microwave simulators are beginning to offer the user a shadow model option [8], with efficiencies reported to be greater than 1000 for nonlinear circuit simulation.

After carefully examining the CV technique, one is forced to make the observation: if we can make a shadow model to circuit simulator correlation coefficient sufficiently high, why don't we instead just perform all of our statistical design operations on the shadow model alone? The large influx of recent papers into the literature indicates that this is the current direction many researchers are taking [9, 10–13]. More will be said on this subject in Chapters 5 and 7.

3.3.3 Importance Sampling

Importance sampling [14] is another very useful concept associated with Monte Carlo yield analysis and yield-estimate variance reduction. To introduce the concept of importance sampling, first consider the following yield-calculation problem. In this problem we wish to derive the yield for a joint parameter PDF, $f_{\mathscr{P}_2}(P)$, from a Monte Carlo yield calculation that used a different joint parameter PDF, $f_{\mathscr{P}_1}(P)$.

Let Y_1 be the yield when parameter density $f_{\mathscr{P}_1}(P)$ is used, and Y_2 be the yield when $f_{\mathscr{P}_2}(P)$ is used such that

$$Y_1 = \int_{-\infty}^{\infty} \text{accept}(P) f_{\mathscr{P}_1}(P) dP$$

$$Y_2 = \int_{-\infty}^{\infty} \text{accept}(P) f_{\mathscr{P}_2}(P) dP$$

In general, Y_2 does not equal Y_1 except when $f_{\mathscr{P}_1}(P) = f_{\mathscr{P}_2}(P)$. To calculate Y_2 from the calculations used for Y_1, rewrite Y_2 as

$$Y_2 = \int_{-\infty}^{\infty} \text{accept}(P) \frac{f_{\mathscr{P}_2}(P)}{f_{\mathscr{P}_1}(P)} f_1(P) dP$$

Then, applying the Fundamental Theorem of Monte Carlo, we get a Monte Carlo estimate for Y_2 as

$$\hat{Y}_2 = \frac{1}{N} \sum_{i=1}^{N} \text{accept}(P_i) \frac{f_{\mathscr{P}_2}(P_i)}{f_{\mathscr{P}_1}(P_i)}$$

where the P_is are chosen according to $f_{\mathscr{P}_1}(P)$ instead of $f_{\mathscr{P}_2}(P)$. This transformation allows one to determine Y_2 by taking samples from $f_{\mathscr{P}_1}(P)$ instead of $f_{\mathscr{P}_2}(P)$.

If we examine the above two forms for the yield estimate of Y_1 and Y_2, we see a relationship that can be exploited. If we save the P_is (chosen according to $f_{\mathscr{P}_1}(P)$) and their corresponding accept(P_i)s when we calculate the yield estimate for Y_1, $\hat{Y}_1$, we can simply calculate the yield, using density $f_{\mathscr{P}_2}$, by scaling each of the accept(P_i)s by the ratio $f_{\mathscr{P}_2}(P_i)/f_{\mathscr{P}_1}(P_i)$. Computationally, this scaling step is negligible.

Another way of looking at the above result is that we have two methods of determining the yield estimate of Y_2, $\hat{Y}_2$: method a,

$$\hat{Y}_2 = \frac{1}{N}\sum_{i=1}^{N} \text{accept}(P_i)$$

where the P_is are chosen according to $f_{\mathscr{P}_2}(P)$; and method b,

$$\hat{Y}_2 = \frac{1}{N}\sum_{i=1}^{N} \text{accept}(P_i)\frac{f_{\mathscr{P}_2}(P_i)}{f_{\mathscr{P}_1}(P_i)}$$

where the P_is are chosen according to $f_{\mathscr{P}_1}(P)$.

Calculating the yield-estimate variance for the two methods, we see that method a yield-estimate variance is given by

$$\sigma_a^2 = \frac{1}{N}\sum_{i=1}^{N} \text{accept}(P_i)^2 - \frac{Y^2}{N}$$

and method b yield-estimate variance is given by

$$\sigma_b^2 = \frac{1}{N}\sum_{i=1}^{N}\left(\text{accept}(P_i)\frac{f_{\mathscr{P}_1}(P_i)}{f_{\mathscr{P}_2}(P_i)}\right)^2 - \frac{Y^2}{N}$$

For proper choice of $f_{\mathscr{P}_1}(P_i)$, σ_b^2 should be less than σ_a^2. The study and determination of the proper $f_{\mathscr{P}_1}(P_i)$ required to reduce σ_b^2 is called *importance sampling*. It has been shown [1] that the optimal sampling PDF is given by

$$f_{\mathscr{P}1\text{opt}}(P) = \text{Accept}(P)\frac{f_{\mathscr{P}_2}(P)}{Y_2}$$

In fact, this sampling results in a zero-variance estimate. However, in practice this cannot be obtained since accept(P) and Y_2 are not known. However, this form shows that the optimal sampling density should be

- Nonzero only where accept(P) is nonzero;
- Large where $f_{\mathscr{P}_2}(P)$ is large.

The sampling density, $f_{\mathscr{P}1\text{opt}}(P)$, puts most of the samples in the important or useful regions of accept(P); hence the label importance sampling.

Researchers have proposed many methods that can be used to approximate $f_{\mathcal{P}1opt}(P)$. Hocevar [3] pointed out that for high-dimension problems or problems where the yield is low, intuitive choices for $f_{\mathcal{P}1opt}(P)$ may not result in variance reduction. Thus, simple attempts to place sample points in the "important" regions may give misleading results. At times this problem has not been properly explained in relation to its utility in the design literature. In high-dimension and low-yield cases, good results are obtained using importance sampling only when close attention is given to the form of the sampling density.

3.4 GEOMETRIC APPROACH TO YIELD CALCULATION

Recall from Section 2.5 that the geometric approach to yield calculation developed yield as the probability that the parameter choice made during manufacturing lies within the acceptability region of parameter space. This idea leads to the understanding that yield is an area (or volume for higher dimension problems), in parameter space, under the parameter density, $f_{\mathcal{P}}(P)$, over the acceptability region, $\mathcal{P}_a$. Figure 2.12 illustrates this concept.

To actually calculate yield using the geometric approximation, the parameter density is assumed to be known. With a geometric approximation to the acceptability region, the volume under the parameter PDF within $\mathcal{P}_a$ can be calculated using the geometry given to $\mathcal{P}_a$ and a numerical integration technique. Although a strict geometric technique is possible, we do not know of any researchers that have developed general-purpose algorithms specifically based on this approach. However, for independent and uniformly distributed parameters, the geometric approach could be very practical.

Section 2.5 also presented the case in which the parameters are statistically independent and jointly uniformly distributed. In this case the geometric interpretation of yield is given mathematically as

$$\text{Yield} = \frac{\text{area}(\mathcal{T} \cap \mathcal{P}_a)}{\text{area}(\mathcal{T})}$$

where area(.) takes the area (or volume when the dimension is greater than two) of the enclosed region. Figure 2.16 shows a two-dimensional example of yield calculation for parameters that are jointly uniform and independent. In this case, it is very easy to visualize the yield-optimization problem. If you know the geometric approximation of $\mathcal{P}_a$ and the tolerance region, you can then evaluate the yield expression by calculating the appropriate areas (or volumes). Several researchers have exploited this technique [15, 16].

3.5 THE PARTS OF YIELD CALCULATION

3.5.1 Background

Two approaches to the description of yield have been presented: mathematics and geometry. Sections 3.3 and 3.4 examined how yield is represented mathematically using integral equations or visualized using a pictorial approach. Both descriptions were developed in the context of a practical model for unit reliability, manufacture, and testing. Before describing the yield-calculation methods, we need to formalize the nomenclature used to describe our manufacturing model. This step will serve two purposes:

1. To provide a general, compact mathematical description of yield calculation techniques (even geometrically based ones) which can be easily converted into computer algorithms;
2. To supply the mathematical skills sufficient to understand the majority of the statistical-design literature.

3.5.2 n-Dimensional Geometry and Convex Sets

In Chapter 2 we said that each unit is represented by a collection of parameters that a designer determined before manufacturing began (see Section 2.3.1). In mathematical terms, the ordered sequence of parameters $P = (p_1, p_2,...p_n)$ is an *ordered n-tuple*. The set of all ordered n-tuples is called *n-space* and is denoted by $\mathcal{R}^n$. In a geometric context, each of the numbers which comprise a given n-tuple can be interpreted in one of two ways: (1) as the coordinates to a point, or (2) as the components to a vector. Thus an ordered n-tuple can be thought of either as a point or as a vector; mathematically the distinction is unimportant, and as such we use the terms interchangeably (unless indicated otherwise).

Often we will be concerned with sets in which we have imposed certain conditions on some or all of the components of the vectors in the set. For example, we might want to consider the set of all points with two components such that the second component is always greater than zero:

$$T = \{(p_1,p_2)|(p_2 > 0)\} \tag{3.15}$$

Here, T is the set of all points (p_1,p_2) such that p_2 is greater than zero. The vertical bar separates the information about the set (represented by the curly braces) into two parts. Information to the left of the vertical bar describes the general form of the points in the set, while any special characteristics a point must possess to be a member of the set are given on the right of the vertical bar.

Before going further, we define some additional terms:

1. A *convex set* is defined as a set where the line segment joining any two points in the set, is also in the set. A convex region must be "solid" (no holes or discontinuities), and its boundaries must not "curve into" the set. Figure 3.3(a and b) show sets that are not convex.
2. A *convex hull* can be thought of as some set that has undergone transformation such that its boundaries are "straightened out", and any holes have been filled in (Figure 3.3c and d).
3. A *convex polyhedron* is the convex hull of a finite set of points.

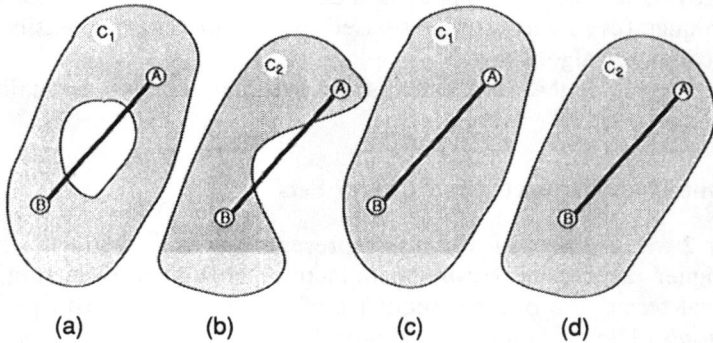

(a)   (b)   (c)   (d)

Figure 3.3 Two nonconvex sets (a) C_1, and (b) C_2 where the line segment AB is not entirely within the set. (c) and (d) are the convex hulls of sets C_1 and C_2 respectively.

3.5.3 The Yield-Calculation Elements

At this juncture in the journey through yield-calculation techniques, we should be able to recognize four basic ingredients necessary for any yield calculation:

1. The unit performance function, $G(.)$;
2. The unit specifications, $S(.)$;
3. The parameter joint probability density function, $f_\mathcal{P}(P)$;
4. Integration over the given region in $\mathcal{P}$.

These elements are illustrated collectively in Figure 3.4.

Yield-calculation methods differ from each other primarily in how $G(.)$ and $f_\mathcal{P}(.)$ are modeled and combined, and how the integration is performed. After examining each of these ingredients in turn, we should be sufficiently equipped to review the literature on yield calculation.

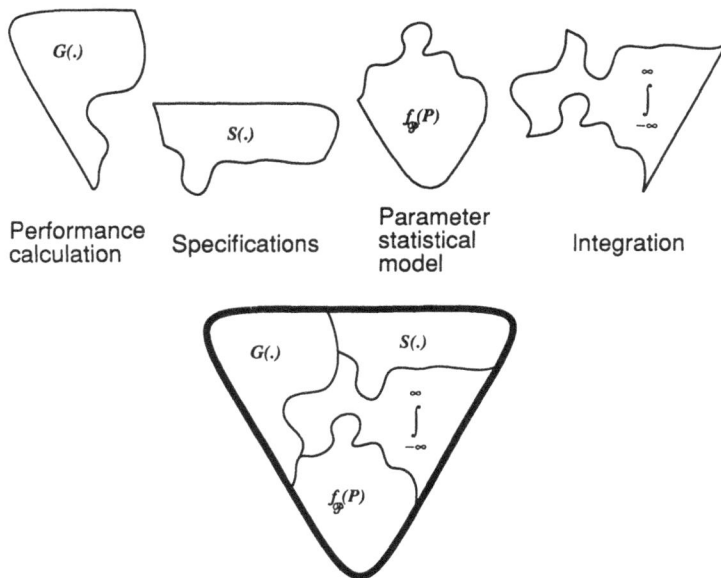

Figure 3.4 The yield-calculation puzzle and its four main parts.

The Unit-Performance Function, G(.)

Consider the vector function G(.) (see Section 2.3.6) which takes unit parameter vectors into unit measurement vectors (i.e., $M_k = G(P_k)$; or $G{:}\mathcal{P}{\Rightarrow}\mathcal{M}$). In electrical and mechanical systems, the transformation G(.) is seldom linear (in P) and typically requires a two-step process. First, a system of equations must be solved. Then the solution vector is further processed to obtain the desired measurements. For example a unit can be characterized in the frequency domain by the complex valued set of simultaneous algebraic equations

$$U(x, P, \omega) = 0 \qquad \omega_1 \le \omega \le \omega_2 \tag{3.16}$$

where x is the vector of node or branch voltages and currents and ω is the frequency. (For units characterized in the time domain, a set of simultaneous differential equations are solved.) Now suppose the measurement of interest is the magnitude of some scattering parameter, S_{mn}. The measurement is determined by further processing of the solution vector [17]. Solving the equation set, U, and determining the measurement is equivalent to evaluating G(P). There are two broad ways of obtaining G(P) in a yield calculation:

1. *Solving the fundamental equations.* This is what is accomplished when a set of parameters is given to a circuit simulator program, and the performance is calculated. We will identify this type of performance calculation by the icon:

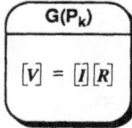

$$G(P_k)$$
$$[V] = [I][R]$$

2. *Approximating the performance relationships.* Because many yield-calculation methods require the repeated evaluation of $G(.)$, it seems natural that researchers would want to find ways to accelerate the process using computationally less expensive models (macro models) and methods. The most popular surrogate for $G(.)$ are low-order (usually quadratic) multidimensional polynomials. The following icon identifies an approximate description of $G(P)$:

$$G(P)$$
$$\approx$$

In Section 3.6, a review of yield-calculation methods will reveal other possibilities for accelerating the evaluation of $G(.)$.

The Unit Specifications, $S(.)$

Simultaneous consideration of unit measurements and specifications gives us the ability to classify a unit as acceptable. Mathematically a unit is acceptable if its performance satisfies a set of unit *constraint* equations

$$C_i = \sum_{j=1}^{k} g_i(x, P, \omega_j) - S_i(\omega_j) \le 0 \quad i = 1, 2, ..., n_c \quad (3.17)$$

where n_c is the number of constraints. Usually, $n_c = M$, the number of measurements (we assume the parameter values are within their feasible limits). K is the number of frequency measurements included in each constraint. Note that the constraints $C_i(P)$ depend on P through the vector transformation $G(P) = (g_1, g_2, ... g_M)$.

If the constraint equations are satisfied for all i, then we say the unit is acceptable or $\text{accept}(P_i) = 1$. The union of the points P_i, where $\text{accept}(P_i) = 1$, forms $\mathcal{P}_a$. There are two means of developing $\text{accept}(P)$ in yield calculation:

1. *Developing a geometric description, often approximate, of* $\mathcal{P}_a$. We identify this type of accept(P) with the icon:

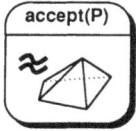

2. *Sampling* accept(P). This is what is performed in a circuit simulator when a unit is simulated with a given set of parameters and the specifications are compared to the performance to determine if the unit passes or fails. This is equivalent to evaluating accept(P_k), for a given P_k. This point-wise investigation of $\mathcal{P}_a$ is identified by the icon:

The Parameter Joint PDF, $f_{\mathcal{P}}(.)$

We have examined parameter statistics and the properties of random variables in Section 2.4. There are two broad methods for modeling $f_{\mathcal{P}}(.)$:

1. Using exact analytical models (that is, equations such as Gaussian, uniform, beta). We will use the following icon to indicate an analytical joint PDF:

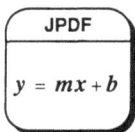

2. Using empirically derived models developed from measured data. We will use the following icon to indicate a measured data joint PDF:

When using an empirical model, which is a common approach, a function can be fit to the data to describe its statistics, or the data can be used directly. Because we deal extensively with statistical modeling techniques in Chapter 6, an in-depth modeling discussion is not given here. In this chapter it will be assumed that the representation for $f_{\mathcal{P}}(P)$ which is required by the method being used is available.

76

Integration

The yield calculation is accomplished through an n-dimensional integration of the product of $f_{\mathscr{P}}(P)$ and accept(P) over the entire parameter space. This often is a high-dimension problem and some clever ways have been developed to handle it. There are at least three ways by which integration is accomplished:

1. *Analysis.* Sometimes enough is known about the integrand to analytically calculate the result directly. For instance if $f_{\mathscr{P}}(P)$ is Gaussian, then a table lookup can determine its area. This is also possible when $f_{\mathscr{P}}(P)$ is approximated by a polynomial in P. Integration by analysis will be represented by the following icon:

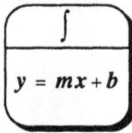

2. *Geometry.* Sometimes the geometry of the acceptability region is known well enough that its weighted volume can be calculated directly from the geometry. This is particularly possible when $f_{\mathscr{P}}(P)$ is uniformly distributed (i.e., it is either a constant or zero as a function of P) and the acceptability region is given a convenient geometric approximation. Integration by geometry will be represented by the icon:

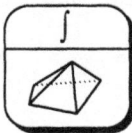

3. *Monte Carlo Integration.* As already stated, the fundamental theorem of Monte Carlo can be used to estimate the value of the yield integral. Most methods use Monte Carlo integration to calculate the yield. Integration by Monte Carlo is represented by the icon:

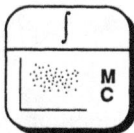

The next section develops the different methods of yield calculation in terms of the elements that we have just presented. Figure 3.5 reviews the icons introduced above.

The unit performance function - $G(.)$

| $G(P_k)$ $[V] = [I][R]$ | Analytically modeled simulation $G(.)$, i.e., Circuit Simulator | $G(P)$ $\approx$ | Approximated simulation model for $G(.)$ |

The unit specifications - $S(.)$

| accept(P_k) $S(.) - G(P_k)$ point-wise evaluation | Analytically modeled simulation $G(.)$, i.e., Circuit Simulator | accept(P) $\approx$ | Create geometrical approximation to accept(P) |

The unit parameter joint PDF - $f(.)$

| JPDF | Sampled data model for the joint PDF | JPDF $f_\oplus(P) \approx$ $N(m, \sigma^2)$ | Analytical model for the joint PDF |

The integration detail

| $\int$ | Integration using a geometrical model | $\int$ $\int x\,dx = \dfrac{x^2}{2}$ | Integration using an analytical model |

| $\int$ M C | Integration using Monte Carlo technique | | |

Figure 3.5 Icons used to classify yield-calculation methods.

3.6 COMBINING THE PARTS: YIELD-CALCULATION METHODS

This section examines how the different elements of yield calculation are combined to form the popular yield-calculation methods given in the literature today.

3.6.1 Regionalization

The regionalization approach involves dividing the parameter tolerance region into many smaller, nonoverlapping regions (cells) to form a n-dimensional rectangular grid. Simulated unit performance at the *center point* of each grid cell c_i, is taken to be the unit performance over the entire grid cell. Using unit specifications, each cell can be classified as a pass region (1), or a failure region (0) (see Figure 3.6).

Once "gridding" is complete, the next step involves preprocessing a sample (for example, 1000 outcomes) from the parameter joint PDF, such that suitable weights (W) for each cell can be calculated:

$$W_i = n_i/N \tag{3.18}$$

where n_i is the number of outcomes within the ith cell, and N is the sample size. These weights represent an approximation to the integral of the joint PDF over the ith grid cell. Then, to economize computational effort, unit-performance evaluations are performed only for those cells with weights above a certain threshold value (so-called "qualified" cells). The main idea here is to trade valuable CPU time for a (hopefully) negligible effect on the accuracy of the yield estimate. There are two reasons why a cell weight would be (relatively) small. First, either the integral of the joint PDF over a given cell is small, or second, the cell volume is proportionality small. After simulating and classifying the weight-qualified cells, the yield is estimated using:

$$\hat{Y} = \sum_{i=1}^{N_q} accept(C_i)w_i$$

where N_q is the set of cell indices of qualified cells.

While regionalization is robust in that it does not require assumptions about the shape of $\mathcal{P}_a$ or the parameter joint PDF $f_{\mathcal{P}}(P)$, there are two main drawbacks:

1. The curse of dimensionality (i.e., 10 parameters divided into 10 regions requires 10^{10} simulations);
2. There is no standard method for the gridding procedure, so the accuracy of the discrete approximation to performance space is difficult to ascertain. (In fact, this question is typically answered with a Monte Carlo experiment.)

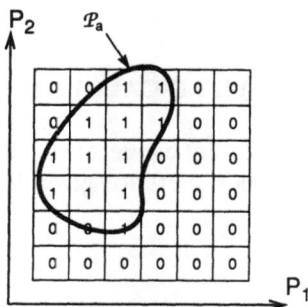

Figure 3.6 Regionalization approximation of the tolerance region by center points of each region.

Spence and Soin [7] summarize by concluding that regionalization does not appear attractive as a practical method for yield estimation. We agree but emphasize that the intuition afforded by studying the regionalization approach is very valuable. A *Yield-Calculation Summary Sheet* for this method is given in Figure 3.7. Summary sheets are given in an identical format for each method discussed.

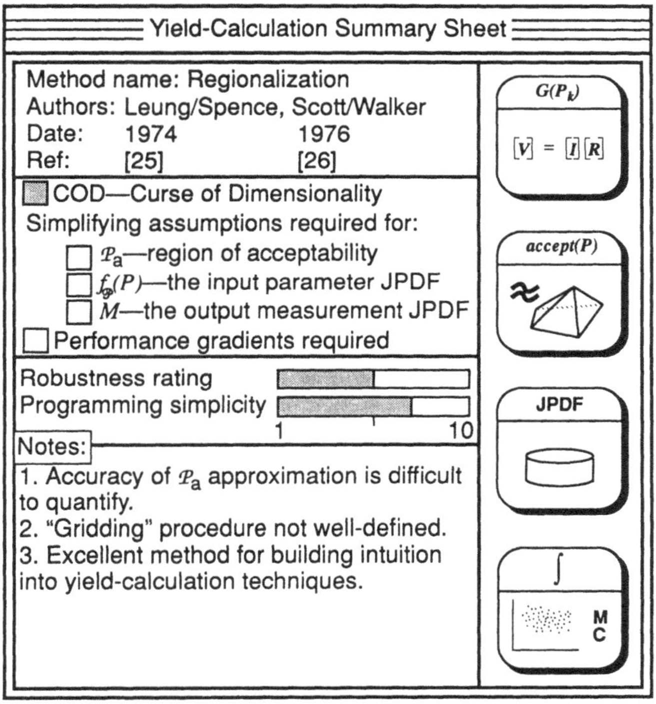

Figure 3.7 Yield-calculation summary sheet using the regionalization method.

3.6.2 Simplical Approximation

Where the regionalizaton approach can be associated with a "discretized" approximation to $\mathcal{P}_a$, the simplical-approximation approach can be associated with a piece-wise linear approximation to the boundary of $\mathcal{P}_a$. Generalized to n dimensions, the approximation to $\mathcal{P}_a$ is built up from hyperplanes (generalized planes) to form a convex polyhedron. Referring to Figure 3.8, the simplical-approximation technique begins by finding a unit which passes all specifications (i.e., accept(P_k) = 1). This step can be accomplished by designer insight or, as is usually the case, a single-point optimizer (see point 1 in Figure 3.8). Next, m $\geq$ n + 1 points on

80

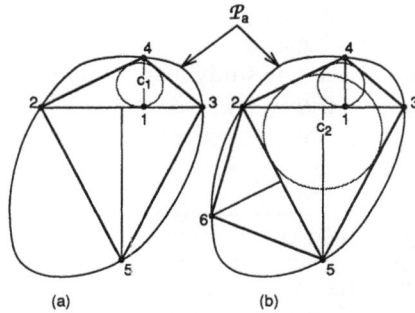

Figure 3.8 Two-dimensional example of the simplical approximation approach: (a) first update, (b) second update.

the boundary of $\mathcal{P}_a$ are found. The m boundary points are found using standard root-finding techniques along lines emanating from the initial pass point. In Figure 3.8, these points are labeled 2, 3, and 4, and were found searching along the parameter axis. The first approximation to $\mathcal{P}_a$ is the "2-d" polyhedron (polygon), made from "2-d" hyperplanes (lines) connecting boundary points 2, 3, and 4. Now that the initial approximation to $\mathcal{P}_a$ is available, repeated application of the following steps are made until satisfactory "convergence" is achieved:

1. Find the center of the largest hypersphere that can be inscribed within the current polyhedron. In our example, the center of a "2-d" hypersphere (circle) is given as the point c_1. In the simplical procedure, this is found using linear programming techniques.
2. Of all hyperplanes "tangent" to the hypersphere found in step 1, determine the one having the largest "face." This hyperplane is identified as the one in which the largest $(n-1)$-dimensional hypersphere may be inscribed. For our example, the tangential 2-d hyperplane "faces" are the line segments 23, 34, and 42. Of these, the largest $(n-1)$-dimensional hypersphere (line) is face 23.
3. Finally, to expand the set of boundary points, a line search is performed in the outward normal direction emanating from the center of the (n_p-1)-dimensional hypersphere found in step 2. Our 2-d example indicates the new member of the boundary point set as 5, and was found along the perpendicular bisector of face 23.
4. Convergence is checked by inscribing a hypersphere within the new polyhedron and examining both the relative and absolute change in the location of the centers of the current and previously inscribed hyperspheres. The algorithm is terminated when these changes are small. Figure 3.8(a) shows the second approximation to $\mathcal{P}_a$ as the polyhedron with faces 25, 53, 34, and 42. Figure 3.8(b) shows the hypersphere center (point c_2) for the second approximation as well as the third approximation to $\mathcal{P}_a$.

While at first this procedure seems tedious, it is actually quite mechanical and straightforward. Once a suitable approximation to $\mathcal{P}_a$ is available, the yield estimate is obtained using Monte Carlo techniques. The rational behind this method hinges on the hope that the computational effort required to build the model, and subsequently use Monte Carlo on it, is less than that required for "plain" Monte Carlo analysis using the unit simulator, G(P).

The simplical-approximation technique is analytically appealing; however, it does require that $\mathcal{P}_a$ be convex. Figure 3.9 shows examples of nonconvex $\mathcal{P}_a$s (additionally, Figure 3.9 (b) is not connected). These $\mathcal{P}_a$s are common among filters as well as other types of circuits [18]. Note that even if the convexity assumption is valid, there is no simple way to quantify the error in the approximation to $\mathcal{P}_a$, except perhaps by using Monte Carlo analysis. Sadly, simplical approximation is yet another approach that suffers from the curse of dimensionality. As a practical method, one should consider using simplical approximation only when the number of design parameters is roughly less than eight [2].

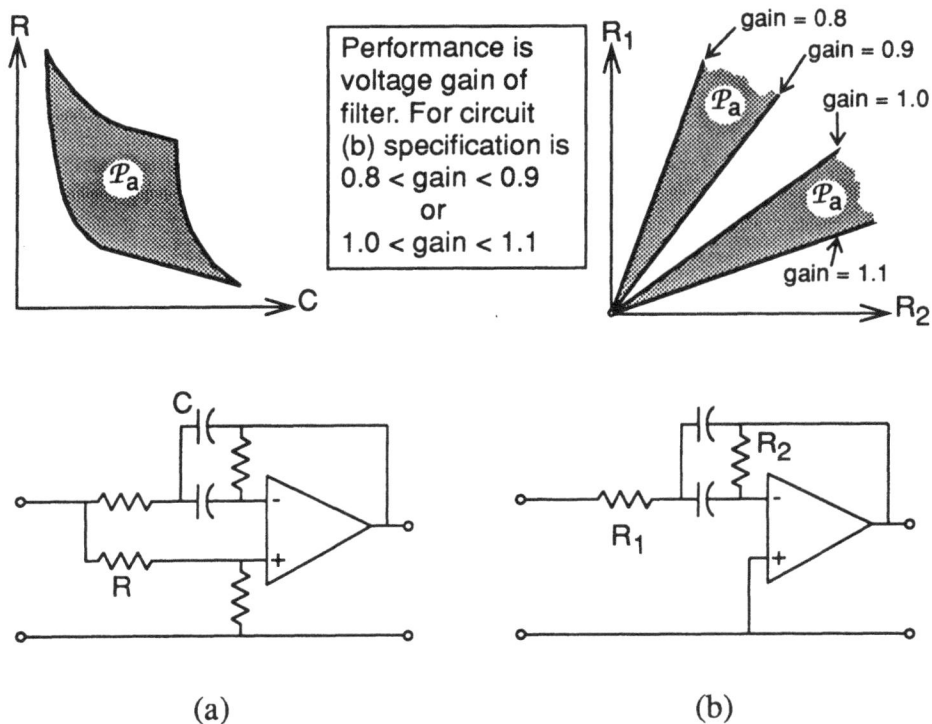

Figure 3.9 Examples of (a) nonconvex $\mathcal{P}_a$ and (b) nonconnected $\mathcal{P}_a$.

Figure 3.10 shows the simplical-approximation summary sheet.

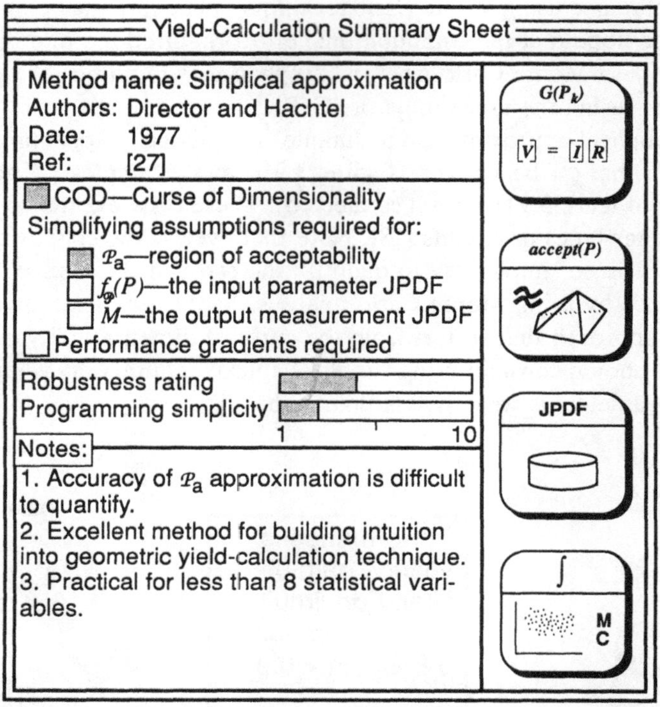

Figure 3.10 Yield-calculation summary sheet using the simplical approximation method.

3.6.3 Efficient Simplical Approximation

To sidestep the heavy computational burden of the original simplical-approximation method, Director, Hachtel, and Vidigal [19] devised a scheme to "progressively" generate the simplical model for $\mathcal{P}_a$. For two relatively simple examples, the authors demonstrated that the new algorithm for building the polyhedron boosts the efficiency of yield-estimate calculations beyond that of standard Monte Carlo, which uses the unit simulator. A simplified flow diagram in Figure 3.11 shows how the algorithm works. First, an initial approximation (polyhedron) to $\mathcal{P}_a$ is built from $n_p + 1$ sample points where $accept(P_k) = 1$. Note that the unit simulator must be used to obtain this approximation, but without the expensive line searches required by the original simplical method. Next, the Monte Carlo yield estimate is obtained by interrogating first the surrogate approximation to $\mathcal{P}_a$. If the sample point falls within the current approximation, it is deemed a pass and the next trial is evaluated. However, if this sample point falls outside the current polyhedron, the unit sim-

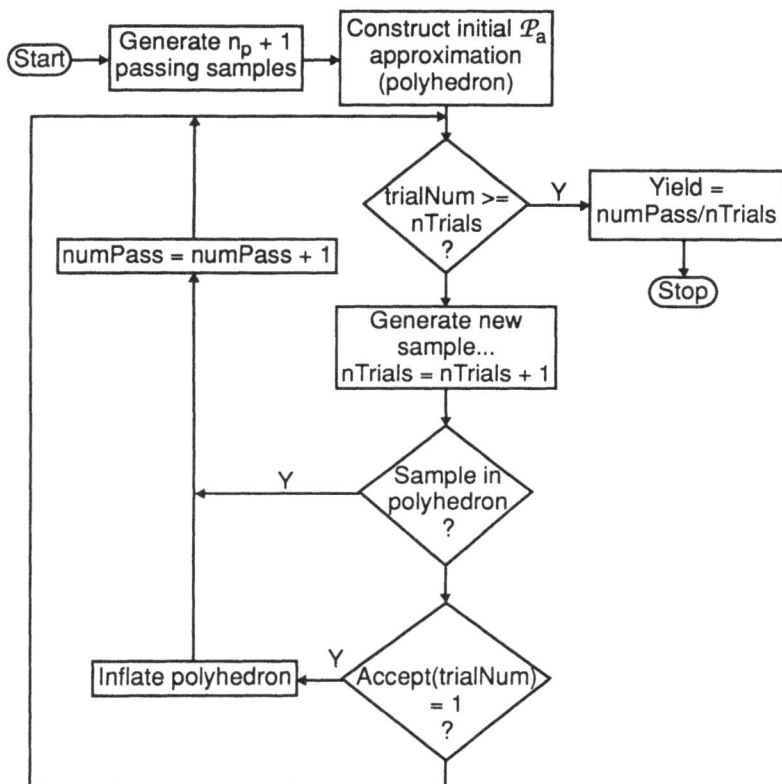

Figure 3.11 Simplified flowchart of the progressive yield-estimation procedure using the simplical approximation to $\mathcal{P}_a$.

ulator is invoked to determine the status of accept(trialNum). If accept(trialNum) = 1 the simplical approximation is inflated to include that point. This process continues until all trials (nTrials) have been evaluated.

One powerful but subtle aspect of a progressive approach to the approximation of $\mathcal{P}_a$ is the fact that the model building effort is expended only over the region in parameter space where we are most likely to require accuracy. Why require that the approximation to $\mathcal{P}_a$ be accurate in places where it is seldom interrogated? Allowing the unit parameter joint PDF to drive the model building effort can be extremely beneficial to computational efficiency.

The downside of this idea is that if the joint PDF changes, say during yield optimization, the accuracy of the surrogate model must be examined and possibly updated.

Figure 3.12 shows the revised-simplical-approximation summary sheet.

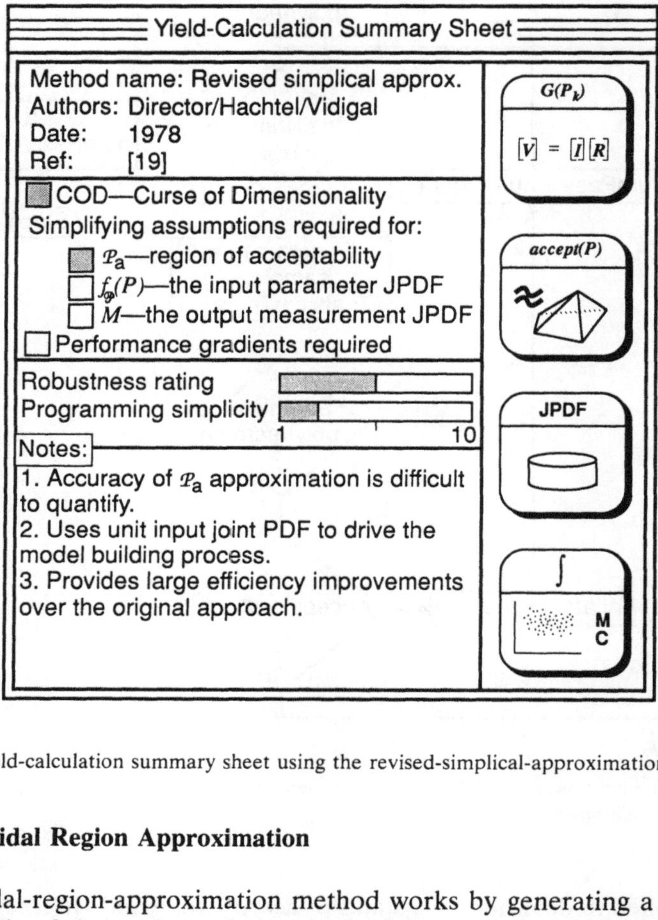

```
╔══════════════ Yield-Calculation Summary Sheet ══════════════╗
║ ┌────────────────────────────────────────┐  ┌──────────┐ ║
║ │ Method name: Revised simplical approx.   │  │  G(Pₖ)   │ ║
║ │ Authors: Director/Hachtel/Vidigal        │  │          │ ║
║ │ Date:     1978                           │  │ [V] = [I][R] │
║ │ Ref:      [19]                           │  └──────────┘ ║
```

Method name: Revised simplical approx.
Authors: Director/Hachtel/Vidigal
Date: 1978
Ref: [19]

■ COD—Curse of Dimensionality
Simplifying assumptions required for:
  ■ $\mathcal{P}_a$—region of acceptability
  ☐ $f_\varphi(P)$—the input parameter JPDF
  ☐ M—the output measurement JPDF
☐ Performance gradients required

Robustness rating
Programming simplicity
    1   10

Notes:
1. Accuracy of $\mathcal{P}_a$ approximation is difficult to quantify.
2. Uses unit input joint PDF to drive the model building process.
3. Provides large efficiency improvements over the original approach.

$G(P_k)$

$[V] = [I][R]$

$accept(P)$

JPDF

$\int$

M
C

Figure 3.12 Yield-calculation summary sheet using the revised-simplical-approximation method.

3.6.4 Ellipsoidal Region Approximation

The ellipsoidal-region-approximation method works by generating a sequence of hyperellipsoids of decreasing volume. Assuming convexity of $\mathcal{P}_a$, the method insures that successive approximations contain $\mathcal{P}_a$ by identifying boundary points on its surface. A hyperplane tangent to $\mathcal{P}_a$ at an identified boundary point is used to divide the space into two regions; one containing $\mathcal{P}_a$ and one not. This information is used to obtain a new hyperellipsoid having smaller volume while still containing $\mathcal{P}_a$ (Figure 3.13). This procedure is repeated until a prespecified volume-reduction ratio is achieved. Yield is then estimated via Monte Carlo techniques using the approximation to $\mathcal{P}_a$ to classify outcomes as pass or fail.

Abdel-Malek exercised the method using two examples. The first, a two-dimensional numerical example (six-sided polygon), and the second, a highpass filter from [20]. For the highpass filter, yield estimates were somewhat less than satisfactory. Using a 500-trial Monte Carlo estimate, the approximation to $\mathcal{P}_a$ gave

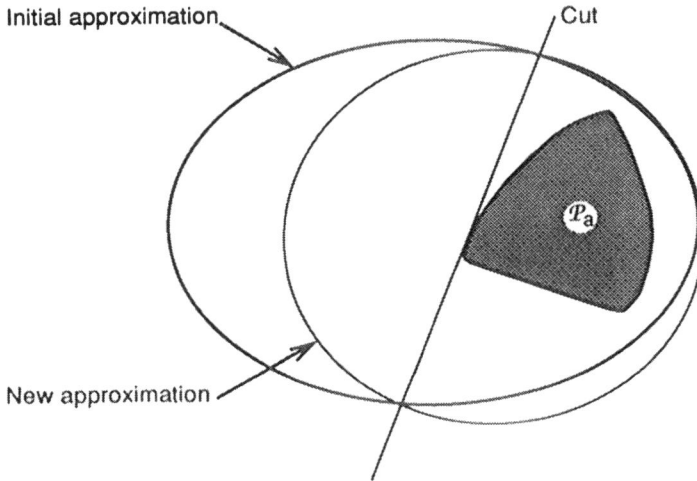

Figure 3.13 Ellipsoidal-region-approximation technique.

a yield of 65.4% and the unit simulator gave a yield of 85.8%. Note that this same example was tested in the revised-simplical-approximation method with excellent results, indicating that the region is most likely convex. Nonetheless the ellipsoidal-region-approximation technique is both visual and intuitive.

Figure 3.14 shows the ellipsoidal-region-approximation summary sheet.

3.6.5 Radial Approximation

The radial-approximation approach to yield estimation is another geometry-based method where yield is computed as the weighted ratio of two volumes. For uniform and uncorrelated unit input parameters

$$Y = V_F/V_T \qquad (3.19)$$

where V_F is the volume of the intersection of the tolerance and acceptable regions, and V_T is the volume of tolerance region. Based on extensive empirical results, Soin and Spence propose an approximation to (3.19) where randomly oriented "lines" emanating from the nominal point are used in the following calculation:

$$\hat{Y} = \frac{1}{2L}\sum_{j=1}^{L}(r_{oj}^+)^{n_p} + (r_{oj}^-)^{n_p}$$

Here, L is the number of "radial lines", n_p is the number of unit parameters (the dimensionality of $\mathcal{P}_a$), and r_{oj}^+ and r_{oj}^- are normalized Euclidean distances associated

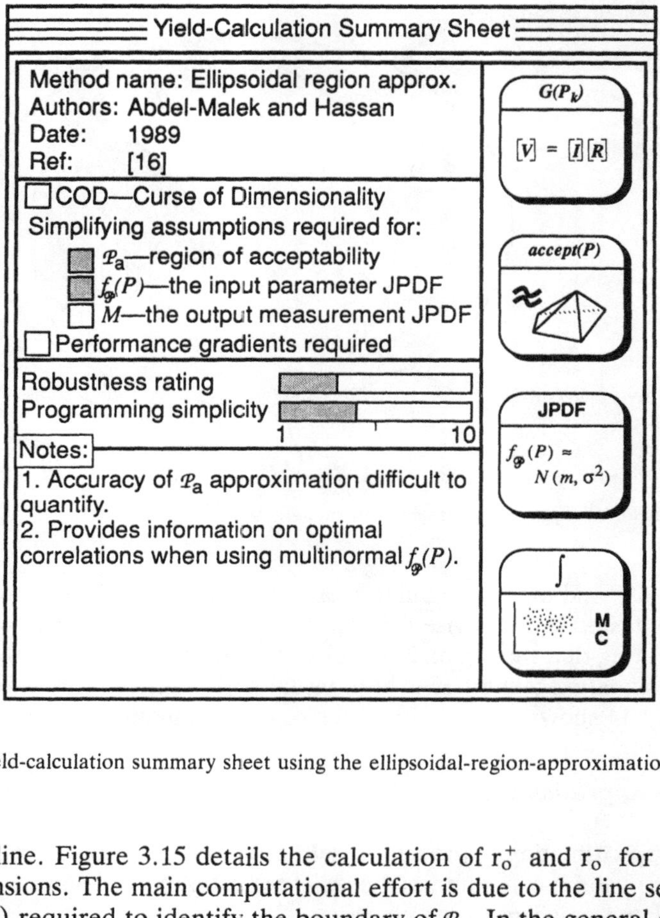

Figure 3.14 Yield-calculation summary sheet using the ellipsoidal-region-approximation method.

with the jth line. Figure 3.15 details the calculation of r_o^+ and r_o^- for a single line in two dimensions. The main computational effort is due to the line searches (two for each line) required to identify the boundary of $\mathcal{P}_a$. In the general case this can involve several unit simulations per line. (For the linear unit-simulation case, there are several shortcuts for economizing CPU expenditure [21].) About 50 lines appear to give satisfactory and consistent results for most problems.

Soin and Spence showed impressive results in accuracy and efficiency, even with an example having 57 random parameters. Such excellent empirical results imply that the radial-approximation approach may not be overly sensitive to nonconvex acceptability regions, $\mathcal{P}_a$.

Figure 3.16 shows the radial-approximation summary sheet.

3.6.6 Polynomial Approximation With Cuts

The polynomial-approximation-with-cuts method, among other features, uses a geometrically based approach to yield estimation. But unlike other approaches,

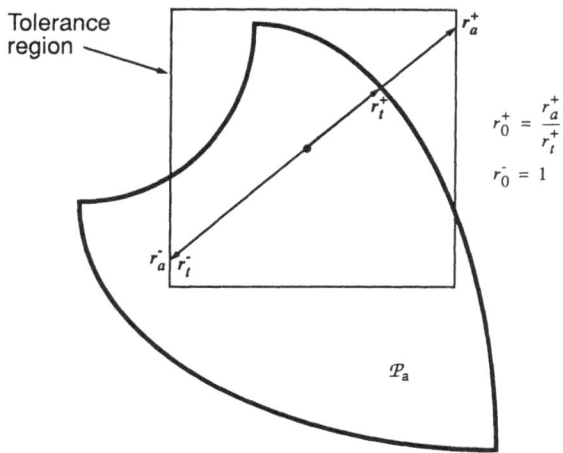

Figure 3.15 The radial-approximation approach.

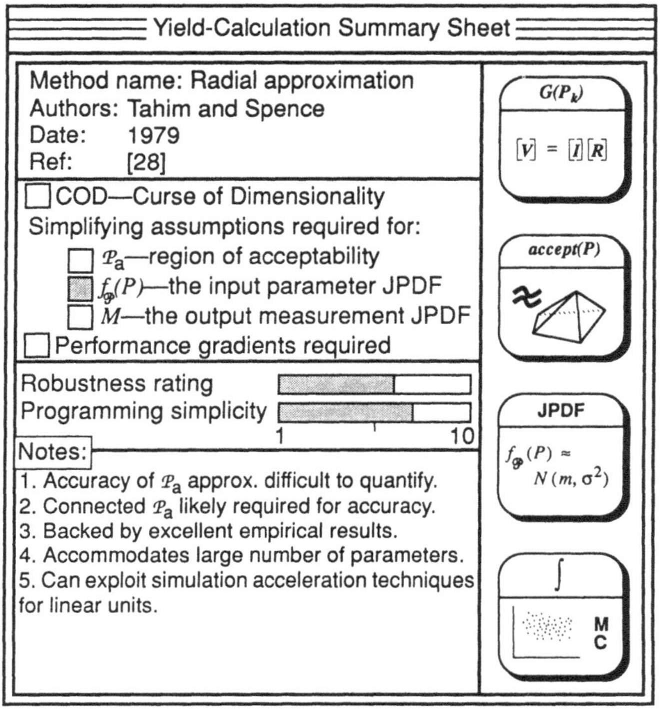

Figure 3.16 Yield-calculation summary sheet using the radial approximation method.

the authors derive explicit formulas for the yield with respect to the unit parameters. These formulas are based upon determination of the weighted volume of the so-called "nonfeasible" region. The nonfeasible region can be identified as that region which is both inside the tolerance orthotope (box), but outside the region of acceptability, $\mathcal{P}_a$. Figure 3.17 illustrates three separate nonfeasible regions based on the linear "cuts" A, B, and C. In the general case, a linear cut can be thought of as a hyperplane where the spatial orientation is determined by the intersection of certain tolerance-box "edges" with the boundary of $\mathcal{P}_a$. Viewed in parameter space, each constraint equation represents a contour of constant performance equal to the specification (the lines C1-C4 of Figure 3.17). Because constraint equations are rarely available in practice, the unit simulator is used to interrogate the performance over certain "critical regions." Then, using the point-wise simulation data, multidimensional quadratic approximations to unit performance as functions of design parameters are obtained. Usually this step requires the solution to a set of simultaneous linear equations. But due to a "fixed pattern" of sample points, the authors are able to derive closed-form expressions for the coefficients of the interpolating polynomials. Finally, based on expressions for the weighted volume of the nonfeasible region, an exact equation for yield with respect to unit parameters

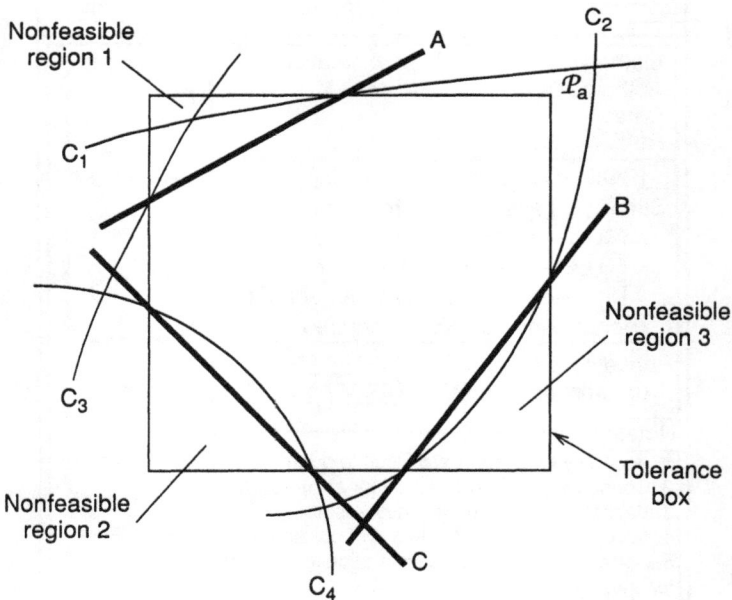

Figure 3.17 Three separate nonfeasible areas based on the linear cuts A, B, and C.

is derived. The simplified formula based on the hypervolume due to each non-feasible region has the form

$$Y = 1 - \sum_{\ell=1}^{m} \frac{V_\ell}{V_T} \qquad (3.20)$$

where V_ℓ is the weighted hypervolume of the ℓth nonfeasible region, V_T is the hypervolume of the tolerance region, and m is the number of linear cuts. (Note that weights are obtained in a similar manner to that of the regionalization approach; see Section 3.6.1.)

While this approach is analytically satisfying, the implementation details of the cuts method are numerous. And not unlike other methods for yield estimation, the cuts technique is not without its own set of advantages, assumptions, and practical considerations. For example, the method begins with the solution to the so-called optimal tolerance-assignment problem (i.e., worst case design where yield = 100% with the smallest possible tolerances). In geometric terms, this problem translates into finding the position and size of the largest tolerance orthotope (box) to fit entirely within $\mathcal{P}_a$. To make the problem tractable, the authors assume that $\mathcal{P}_a$ is *one-dimensional convex*. This assumption allows for $\mathcal{P}_a$ to be nonconvex overall, but with the requirement that if the vertices of a tolerance orthotope are within $\mathcal{P}_a$, then the entire tolerance orthotope must be within $\mathcal{P}_a$ (see Figure 3.18). While somewhat less restrictive than the region assumption in the simplical-approximation approach, one-dimensional convexity is nonetheless a requirement for which we cannot easily guarantee conformity.

The curse of dimensionality can also become a factor in the cuts method. Recall that polynomial constraint approximations are built and refined in regions

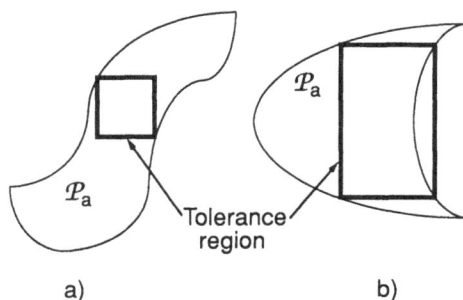

Figure 3.18 Two nonconvex regions of acceptability $\mathcal{P}_a$, where (a) is also one-dimensional convex, but (b) is not.

of "critical importance." These regions are defined by the set of tolerance-box vertices that are very near to, or on the boundary of $\mathcal{P}_a$—the so-called active vertices. If you assume that half of the vertices will be "active," then there will be $2^{(n_p-1)}$ interpolations, each requiring $(n_p+1)(n_p+2)/2$ unit simulations, where n_p is the number of unit design parameters.

Several simple circuit examples from [15] and [22] illustrate the accuracy and utility of the method. A more involved example using a current switch emitter follower (CSEF) circuit was the focus of [23]. The example presents several of the cut method's advantages:

1. Only 105 unit simulations were required (using a single interpolation region).
2. By having approximations to the constraint equations, no additional circuit simulations are necessary when considering alternative specifications.
3. The optimal tolerance problem is solved providing the designer with the "cheapest" solution to the 100% yield problem.

Figure 3.19 shows the summary sheet of the polynomial approximation with cuts.

3.6.7 Dynamic Constraint Approximation

The focus of the dynamic-constraint-approximation approach is to substantially reduce the number of unit simulations used in statistical design and optimization procedures. By formulating an efficient method for characterizing a unit's performance with respect to its design parameters, the authors have contributed a tool from which all statistical design and optimization theorists can benefit.

The basis of the approach is to approximate unit performance functions by multidimensional interpolating polynomials. For a single unit-performance measurement, the general form for the interpolating polynomial is given as

$$\hat{g}(P) = a_0 + \sum_{i=1}^{np} a_i p_i + \sum_{\substack{i,j=1 \\ i \leq j}}^{np} a_{ij} p_i p_j + \sum_{\substack{i,j,k=1 \\ i \leq j \leq k}}^{np} a_{ijk} p_i p_j p_k + K \cong g(P) \quad (3.21)$$

where P is the general unit parameter vector and the a's are the unknown coefficients of the interpolating polynomial. We write $\hat{g}(P)$ because this is an approximate mapping function relating the unit input parameter vector P to a unit performance vector M. In order to solve (3.21) for the unknown coefficients, the actual unit simulator, G(P) must be invoked as many times as there are unknowns.

In practical situations, only the first three terms of (3.21) are used. However, suppose that we choose the linear model, where only the first two terms of (3.21)

```
╔══════════ Yield-Calculation Summary Sheet ══════════╗
║ Method name: Polynomial approx. w/ cuts  │  ╭─────────╮ ║
║ Authors: Bandler and Abdel-Malek         │  │  G(P)   │ ║
║ Date:    1978                            │  │         │ ║
║ Ref:     [15]                            │  │   ≈     │ ║
║ ──────────────────────────────────────  │  ╰─────────╯ ║
║ ▣ COD—Curse of Dimensionality            │             ║
║ Simplifying assumptions required for:    │             ║
║    ▣ 𝒫a—region of acceptability          │  ╭─────────╮ ║
║    ▣ f𝒫(P)—the input parameter JPDF      │  │accept(P)│ ║
║    ☐ M—the output measurement JPDF       │  │  ≈  ◁   │ ║
║ ☐ Performance gradients required         │  ╰─────────╯ ║
║ ──────────────────────────────────────  │             ║
║ Robustness rating      [▓▓▓▓▓ ]          │             ║
║ Programming simplicity [▓     ]          │  ╭─────────╮ ║
║                        1        10       │  │  JPDF   │ ║
║ ──────────────────────────────────────  │  │f𝒫(P) =  │ ║
║ Notes:                                   │  │N(m,σ²)  │ ║
║ 1. Accuracy of 𝒫a approximation difficult to│ ╰────────╯ ║
║ quantify.                                │             ║
║ 2. Approximate yield sensitivities available.│         ║
║ 3. Excellent theoretical development.    │  ╭─────────╮ ║
║ 4. Monte Carlo integration can be used to│  │   ∫     │ ║
║ accommodate arbitrary input JPDF's.      │  │   ◁     │ ║
║ 5. One-dimensional convexity assumed.    │  ╰─────────╯ ║
╚══════════════════════════════════════════════════════╝
```

Figure 3.19 Yield-calculation summary sheet using the polynomial approximation method with cuts.

are retained, hence the number of unknown coefficients a_i is $n_p + 1$. To solve the system, we invoke the unit simulator using $n_p + 1$ different values of P_k, commonly referred to as "base points." Finally, a system of linear equations is solved to find the coefficients a_i of our linear model.

But rarely will the linear model provide us with enough accuracy. If we use the quadratic model, another $n(n+1)/2$ base points (unit simulations) will be required; perhaps an excessive computational cost when the overall optimization algorithm is considered. But any fewer base points and the resulting system of equations becomes underdetermined (i.e., there are an infinite number of solutions with respect to the coefficients, such that no unique quadratic interpolation can be obtained).

To solve this problem and force uniqueness with a base-point count between $n_p + 1$ and $(n_p + 1)(n_p + 2)/2$, the authors impose a "maximally flat" rule. Analytically, this amounts to choosing the second-order and mixed-term coefficients to be as small as possible in the least squared sense. This approach is justified by the fact that since we really do not know the nature of the nonlinearity of G(P) it is "safe" to assume that it is as close to linear as possible. They refer to this type of interpolation as "least prejudiced" and formulate the following quadratic programming problem:

$$\text{minimize } \|v\|^2$$
$$v$$
$$\text{subject to } Qz = b$$

where V is the vector of second order and mixed term coefficients, Q is the matrix of base points, and b is the vector of responses corresponding to the base points. Additionally, the authors provide an updating scheme which requires only a few simple matrix operations any time a new base point is added, thus eliminating the need to resolve the entire system of equations.

The suitability of the approach was tested on a nine-element filter which was used elsewhere to compare alternate methods for design centering [24]. The results they obtained were somewhat astounding in that with the addition of between one and five base points beyond that required for the linear model, prediction accuracy went from inadequate to excellent for both response and yield calculations.

Biernacki [13] extended the original maximally flat quadratic approach by combining it with a fixed pattern of base points (see Section 3.6.6). The new formulation allows for reduced memory and computational requirements, in addition to simplicity of implementation, but at the expense of model accuracy. The result of the fixed pattern of base points when applied to the maximally flat interpolation is that all mixed-term coefficients are forced to zero. Additionally, the number of base points required for a quadratic model falls between $n_p + 1$ and $2n_p + 1$.

Good results were reported with the new modeling approach where, combined with Monte Carlo yield estimation, several medium-scale circuit problems were examined. One example had 50 random unit parameters, while another with 11 showed relative efficiency improvements over the original approach of about 52%.

Figure 3.20 shows the dynamic-constraint-approximation summary sheet.

3.6.8 Monte Carlo

Without debate, the easiest, most forgiving method to implement for yield estimation has to be the Monte Carlo method. You will usually find Monte Carlo at

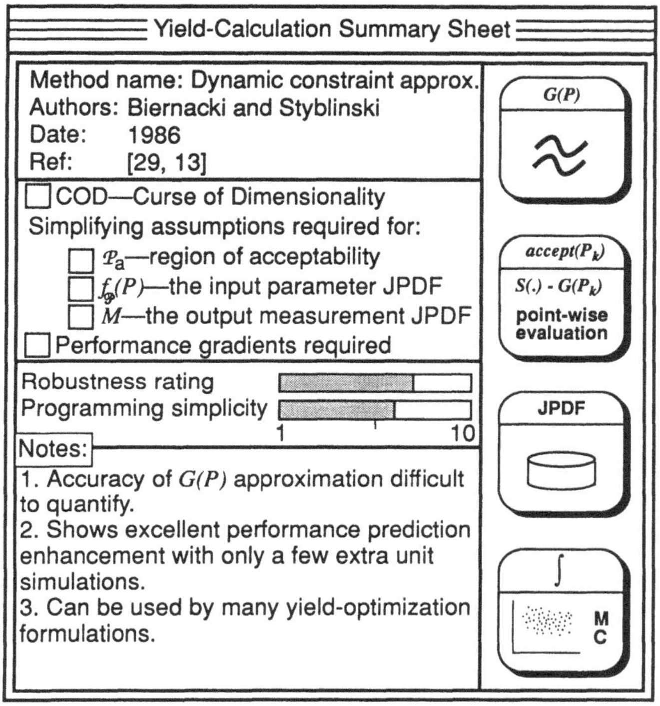

```
╔══════════ Yield-Calculation Summary Sheet ══════════╗
║ Method name: Dynamic constraint approx.  │  G(P)    ║
║ Authors: Biernacki and Styblinski        │          ║
║ Date:    1986                            │   ≈      ║
║ Ref:     [29, 13]                        │          ║
║ ☐ COD—Curse of Dimensionality                       ║
║ Simplifying assumptions required for:    │ accept(Pk)║
║   ☐ Pa—region of acceptability           │ S(.) - G(Pk)║
║   ☐ fg(P)—the input parameter JPDF       │ point-wise║
║   ☐ M—the output measurement JPDF        │ evaluation║
║ ☐ Performance gradients required                    ║
║ Robustness rating      [████████  ]                 ║
║ Programming simplicity [███████   ]      │  JPDF    ║
║                        1          10                ║
║ Notes:                                              ║
║ 1. Accuracy of G(P) approximation difficult         ║
║ to quantify.                                        ║
║ 2. Shows excellent performance prediction │  ∫      ║
║ enhancement with only a few extra unit    │      M  ║
║ simulations.                              │      C  ║
║ 3. Can be used by many yield-optimization           ║
║ formulations.                                       ║
╚═════════════════════════════════════════════════════╝
```

Figure 3.20 Yield-calculation summary sheet using the dynamic-constraint-approximation method.

the root of most practical methods for yield estimation and optimization; it simply fits in anywhere. If its not at the root of a method, then it probably exists as a support or back-up routine; you see this often, for example, in methods for design centering where once a number of iterations are performed, a Monte Carlo yield estimate is obtained to confirm the results of the centering job. The Monte Carlo-algorithm flow diagram is depicted in Figure 3.21. We also thought it fitting to include a Yield-Calculation Summary Sheet, but the authors and date are somewhat nebulous. It is believed that the earliest documented use of random sampling for finding the solution to an integral problem is due to Comte de Buffon in 1777 [1]. Kalos also notes that the name "Monte Carlo" was first used by scientists working on the Manhattan project during the 1940s.

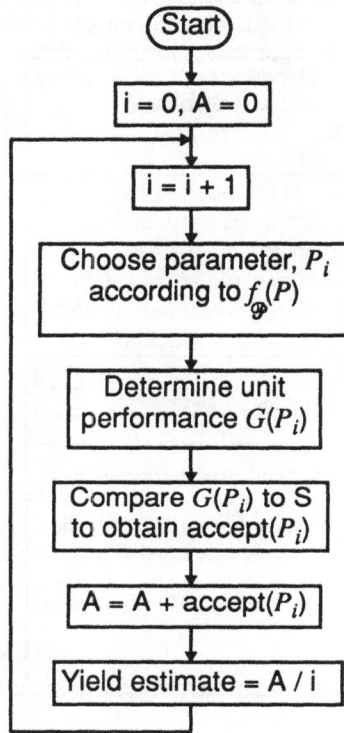

Figure 3.21 Algorithm flow diagram for the Monte Carlo method for calculating yield.

Figure 3.22 shows the Monte Carlo method summary sheet follows.

3.7 CONCLUSION

If we are to "design in" the capability for our products to withstand uncontrollable random parameter variations, then an understanding of yield-calculation methods is necessary. In this chapter we examined various proposals for yield calculation, breaking them down into four principle elements: $G(.)$, $S(.)$, $f_{\mathcal{P}}(.)$, and a method for integrating these elements. Because it is fundamental to almost all methods for calculating yield, we presented a detailed coverage of Monte Carlo methods including variance-reduction techniques. We will revisit most of the yield-calculation techniques covered in this chapter again in Chapter 5, where our concern will be maximizing the yield.

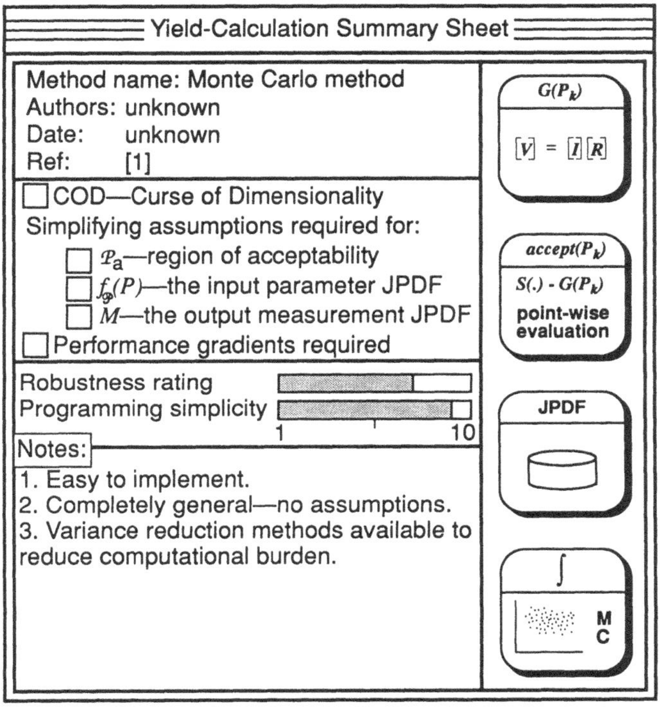

Figure 3.22 Yield-calculation summary sheet using the Monte Carlo method.

3.8 IMPORTANT IDEAS FROM CHAPTER 3

Section 3.1

* Successful design for manufacturability requires that the designer have an understanding about the subtleties of yield calculation.

Section 3.2 & 3.3

* Monte Carlo integration can be regarded as a powerful tool for evaluating complicated multidimensional definite integrals.
* A Monte Carlo integration estimate
  1. Is a random variable having calculable error bounds;
  2. Has predictable statistical accuracy independent of the dimensionality of the integration;
  3. Is easily applied to the yield integral.

- For fixed confidence in the estimate, error in the Monte Carlo estimate goes inversely proportional to the square root of the number of trials.
- Variance-reduction techniques can be used to reduce the error in a Monte Carlo estimate for a fixed number of trials and confidence in the estimate.

Section 3.4

- The geometric approach to yield calculation is limited practically to uniform and independent joint parameter densities.
- Monte Carlo integration methods can be thought of in terms of numerically evaluating the geometric definition for yield (i.e., intersection of weighted volumes in parameter space.)

Section 3.5

- Any given yield-calculation method can be separated into four basic elements:
  1. $G(.)$, the unit performance function;
  2. $S(.)$, the unit specifications;
  3. $f_\mathcal{P}(P)$, the joint probability density function;
  4. Integration of items 1, 2, and 3 over parameter space.
- $G(.)$ is a vector function which maps unit parameter vectors into unit measurement vectors: $G: \mathcal{P} \Rightarrow \mathcal{M}$.
- For most units, $G(P)$ is seldom linear in P, and typically requires the lengthy solution of a system of equations. This is why many methods use approximations to $G(.)$ (macromodels) which do not involve a matrix solve. However, it is usually difficult to determine the extent to which errors in the macromodel will affect the accuracy of the yield estimate.
- Simultaneous consideration of unit measurements M (the result of $G(.)$) and specifications $S(.)$, give rise to a set of vector constraint equations C_i.
- The testing function accept(P_k) is defined in terms of the constraint equations so that a unit can be classified as "pass" or "fail."
- The parameter joint PDF $f_\mathcal{P}(P)$ is modeled in two ways:
  1. Using exact equations (uniform, Gaussian, etc.);
  2. Empirically (using equations derived from sampled data).
- Yield-calculation methods combine $f_\mathcal{P}(P)$ and accept(P) by integration over parameter space $\mathcal{P}$. This integration is accomplished:
  1. Analytically;
  2. Geometrically;
  3. Numerically (i.e., using the Monte Carlo technique).

Section 3.6

- Yield-calculation methods differ in the way they define and combine the four elements of yield calculation.

- Most methods for yield calculation use Monte Carlo techniques.
- The most computationally intensive aspect of yield calculation is usually the evaluation of the unit-performance function G(.). This is why many methods attempt to form some sort of efficient (hopefully accurate) surrogate for G(.), or utilize accelerated Monte Carlo techniques (variance-reduction ideas).

EXERCISES

Exercise 3.1

We suggest that you try this exercise because evaluating integrals as the ratio of volumes and using the fundamental theorem of Monte Carlo can lead to a visualization for solving these kinds of problems by "throwing darts" at regions in n-space.

An example of the use of this technique would be to try to determine the ratio of the area of the state of Idaho to the entire area of the United States. The Monte Carlo trials for this problem will be to uniformly drop potatoes across the entire US. In the trials you would keep track of the number that land in the state of Idaho and the total number dropped. The desired area ratio estimate is:

$$\text{area ratio estimate} = \frac{\text{number of potatoes falling in Idaho}}{\text{total number of potatoes dropped}}$$

This area-ratio estimate is an unbiased estimate. Figure 3.23 shows the areas involved here and serves as a reminder of the complexity of the regional descriptions used in this example. We think that you will see that this technique also

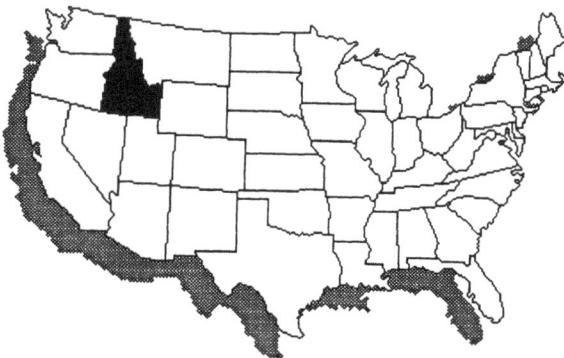

Figure 3.23 US Map showing area ratio that can be determined by using Monte Carlo "trials."

works for Hawaii, which is not a connected state, and for Wisconsin which is not a convex state. Furthermore, to make this trial, an exact map of the US is not needed, just some way of determining if each potato falls inside or outside of Idaho. The generality of these Monte Carlo methods make them powerful tools for evaluating complex definite integral equations, like the yield equations.

REFERENCES

[1] M.K. Kalos and P.A. Whitlock, *Monte Carlo Methods, Volume 1: Basics*, New York NY: John Wiley and Sons, 1986.

[2] R.K. Brayton, G.D. Hachtel, and A.L. Sangiovanni-Vincentelli, "A Survey of Optimization Techniques for Integrated Circuit Design," *Proc. IEEE*, Vol 69, No. 10, pp. 1334-1363, 1981.

[3] D.E. Hocevar, M.R. Lightner, and T.N. Trick, "A Study of Variance Reduction Techniques for Estimating Circuit Yields," *IEEE Trans. on Computer-Aided Design*, Vol. CAD-2, No.3, July, 1983, pp. 180-192.

[4] J.M. Hammersley and D.C. Handscomb, *Monte Carlo Methods*, London England: Methuen, 1964.

[5] T.B. Neill, "Variance Reduction in Monte Carlo Analysis of Electrical Networks," *IEE Conf. on CAD*, April 1972, Inst. Elec. Engr. Publication #86, pp. 219-224.

[6] R.S. Soin and P.J. Rankin, "Efficient Tolerance Analysis Using Control Variates," *IEE Proceedings*, Pt. G, Vol. 132, No. 4, 131-142, 1985.

[7] R. Spence and R. Soin, *Tolerance Design of Electronic Circuits*, Reading MA: Addison-Wesley, 1988.

[8] EEsof Inc., Westlake Village, CA.

[9] T.K. Yu, S.M. Kang, I.N. Hajj, and T.N. Trick, "iEdison: an Interactive Design Tool for MOS VLSI Circuits," *IEEE ICCAD Int. Conf. on CAD*, Nov. 1988, pp 20-23.

[10] T.K. Yu, S.M. Kang, J. Sacks, and W.J. Welch, "An Efficient Method for Parametric Yield Optimization of MOS Integrated Circuits," *IEEE ICCAD Int. Conf. on CAD*, Nov. 1988, pp 190-193.

[11] K.K. Low and S.W. Director, "A New Methodology for the Design Centering of IC Fabrication Processes," *IEEE ICCAD Int. Conf. on CAD*, Nov. 1989, pp. 194-197.

[12] T.K. Yu, S.M. Kang, I.N. Hajj, and T.N. Trick, "Statistical Performance Modeling and Parametric Yield Estimation of MOS VLSI," *IEEE Trans. on CAD*, Vol. CAD-6, no. 6, Nov. 1987, pp 1013-1022.

[13] R.M. Biernacki and J.W. Bandler, "Efficient Quadratic Approximation for Statistical Design," *IEEE Trans. on Circuits and Systems*, Vol CAS-36, No. 11, Nov. 1989, pp. 1449-1454.

[14] K. Singhal and J. F. Pinel, "Statistical Design Centering and Tolerancing Using Parametric Sampling," *IEEE Trans. on Circuits and Systems*, CAS-28, No. 7, pp 692-702, 1981.

[15] J.W. Bandler and H.L. Abdel-Malek, "Optimal Centering, Tolerancing and Yield Determination Via Updated Approximations and Cuts," *IEEE Trans. on Circuits and Systems*, Vol. CAS-25, pp. 853-871, 1978.

[16] H.L. Abdel-Malek, and S.O. Hanson, "The Ellipsoidal Technique for Design Centering and Region Approximation," *IEEE Trans. on CAD*, Vol. CAD-10, No. 8, 1991, pp. 1006-1014.

[17] G. Gonzalez, *Microwave Transistor Amplifiers*, Englewood Cliffs, NJ: Prentice Hall, 1984.

[18] E. Butler, "Realistic Design Using Large-Change Sensitivities and Performance Contours," *IEEE Trans. Circuit Theory*, Vol. CT-18, Jan. 1977, pp.58-66.

[19] S.W. Director, G.D. Hachtel, L.M. Vidigal, "Computationally Efficient Yield Estimation Procedures Based on Simplical Approximation," *IEEE Trans. on Circuits and Systems*, CAS-25, No. 3, pp. 121-130, 1978.

[20] Pinel and Roberts, "Tolerance Assignment in Linear Networks Using Non-linear Programming," *IEEE Trans. on Circuit Theory,* CT-19, No. 5, pp 475-79.

[21] R. Brayton, R. Spence, *Sensitivity and Optimization,* Section 6.7, New York, NY: Elsevier, 1980.

[22] H.L. Abdel-Malek and J.W. Bandler, "Yield Optimization for Arbitrary Statistical Distributions, Part I: Theory," *IEEE Trans. on Circuits and Systems,* Vol. CAS-27, pp. 245-253, 1980.

[23] H.L. Abdel-Malek and J.W. Bandler, "Yield Optimization for Arbitrary Statistical Distributions, Part II: Implementation," *IEEE Trans. on Circuits and Systems,* Vol. CAS-27, pp. 253-262, 1980.

[24] E. Wehrhahn and R. Spence, "The Performance of Some Design Centering Methods," *Proc. of the IEEE Int. Symp. on Circuits and Systems,* Montreal Canada, May 1984, pp. 1424-1438.

[25] K.H. Leung and R. Spence, "Efficient Statistical Circuit Analysis," *Electronic Letters,* Vol 10, pp. 360-362, 1974.

[26] T.R. Scott and T.P. Walker, "Regionalization: A Method for Generating Joint Density Estimates," *IEEE Transactions on Circuits and Systems,* CAS-23, No. 4, pp. 229-234, 1976.

[27] S.W. Director and G.D. Hachtel, "The Simplical Approximation Approach to Design Centering," *IEEE Transactions on Circuits and Systems,* CAS-24, pp. 363-372, 1977.

[28] K.S. Tahim and R. Spence, "A Radical Exploration Approach to Manufacturing Yield Estimation and Design Centering," *IEEE Transactions on Circuits and Systems,* CAS-26, No. 9, pp. 768-774.

[29] R.M. Biernacki and M.A. Styblinski, "Statistical Circuit Design With a Dynamic Constraint Approximation Scheme," *Proc. IEEE Int. Symp. Circuits Syst.,* San Jose, CA, 1986, pp. 976-979.

Chapter 4
Statistical Sensitivity

"One important ingredient of quality is uniformity. It seems almost too simple and insignificant." [1]

"When a product's performance deviates from the target performance, its quality is considered inferior." [2]

"In analyzing variation there are two kinds of mistakes we could make:

1. We could mistake the cause of variation as being special in nature, when in fact it is random and caused by the system (common causes).
2. We could mistake the source of variation as being systematic in nature (common causes), when in fact it is special in nature (a special cause) and can and should be identified and, if possible, eliminated." [1]

4.1 INTRODUCTION

This chapter contains an introduction to the concept of statistical sensitivity. Sensitivities are often used in circuit design and optimization to provide insight into the relationships between unit parameters and performance. There are three main uses of sensitivities [3]:

1. Sensitivities give the designer understanding of how variations in parameters affect performance. This can identify the parameters which have the most influence on performance.

2. Sensitivities can be used to choose among several circuit (or unit) structures that all have the same nominal performance. The general consensus is that the structure with the least performance sensitivity is superior.
3. Sensitivities are used in optimization to determine performance gradients.

There are many different sensitivity definitions, each having its own characteristics. This chapter presents the most popular sensitivities and compares them in two examples, one mechanical and one electrical. Statistical sensitivities, which we propose to use in statistical design and analysis, are also presented.

4.1.1 Classic Sensitivity

The simplest and most used definition, which we call the *classic sensitivity,* is the derivative of the performance, G(P), with respect to some parameter in P, say p_i. This is given as

$$S_{p_i}^G = \frac{\partial G(P)}{\partial p_i}$$

which equals the sensitivity of G(P) with respect to changes in the parameter p_i. This type of sensitivity measure takes a look at the variation of the performance for infinitesimally small changes of a parameter, one parameter at a time.

The most useful forms of the sensitivities are normalized to remove the influence of parameter and performance units on the magnitude of the sensitivity. For instance, a normalized classical sensitivity can be

$$SN_{p_i}^G = \frac{p_i}{G(P)} S_{p_i}^G$$

This normalization gives the sensitivity a simple interpretation. If the normalized sensitivity is 2.0, then a 1% change in the parameter will cause a $+2\%$ change in the performance. Although normalization is recommended in practice, for simplicity of notation and presentation, the normalization is included only when necessary.

4.1.2 Interpretation of Sensitivity

Performance sensitivities are useful because their interpretation can provide insight into the relationships that exist among the unit parameters and performance. For instance, parameters that have small sensitivities are said to have little influence on performance. In a manufacturing context, these parameters may not need to be tightly controlled. Conversely, parameters which have large sensitivities are identified as critical to performance. In manufacturing, these parameters may need to be carefully monitored and controlled or have their sensitivities reduced by design or manufacturing changes.

4.1.3 Sensitivity in Optimization

Optimization techniques have become very useful in improving the performance of most designs. Complex optimization systems can choose among parameter values to minimize a given performance error function. Gradient methods are often used in optimization. Figure 4.1 shows a flow chart for optimization using a gradient method.

In general when using gradient optimization, sensitivity analysis is performed and the results of this analysis drive a parameter-modification routine. The parameter-modification routine chooses new parameter values so that at each step in the optimization, the overall error function is reduced, as described by the sensitivity block. In present day optimizers, the sensitivity is almost always based upon the normalized classic sensitivity. In this chapter we will show that the use of the classic sensitivities in design and optimization can lead to designs with poor manufacturing yield. A recent proposal on the use of statistical sensitivities for use in gradient optimization [4], is also described.

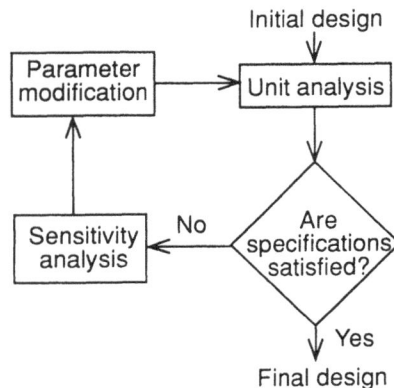

Figure 4.1 The gradient-optimization process.

4.1.4 Manufacturing Sensitivity

One important goal for a manufacturable unit is that the unit's statistical performance, such as performance average value, performance variance, or yield, be acceptable during manufacture. (As noted in Chapter 3, unit service life can easily be incorporated into this formulation.) To better state the conceptual goal for a manufacturing-oriented sensitivity study, we define *manufacturing sensitivity* as the change in a performance statistic during manufacture as a function of the design parameter changes. The unit design parameter values are given by the designer to the manufacturing engineer. In the context of design for manufacturability, the unit design parameters specified by the designer become the nominal parameters used in manufacturing. The manufacturing environment then imposes a statistical joint density around the nominal design parameters, and samples from this joint density are used as parameter values during the manufacture of each unit. (We assume that the parameter joint density can be sufficiently approximated, and that it is not changing with time.) So, a major goal for manufacturing-oriented design is to measure and manage the manufacturing sensitivity of the unit design.

To illustrate the concept of manufacturing sensitivity, consider an amplifier design. The amplifier parameter of interest is a certain bias resistor, R. With the nominal value of R set at 1.0 kΩ, 10,000 amplifiers are manufactured and the average power gain, averaged over the 10,000 units, is found to be 642. Then with the nominal value set at 1.1 kΩ, another 10,000 amplifiers are manufactured and the average power gain is found to be 625. The manufacturing sensitivity of the average power gain to the changes in the bias resistor, R, is approximately (625 − 642) / (1.1 − 1.0 kΩ) = −170/kΩ. This figure indicates that as R increases by 1 kΩ, the average manufactured gain will decrease by 170. The manufacturing sensitivity shows how the parameter affects a performance statistic in the actual physical manufacturing environment.

Unfortunately, manufacturing sensitivity can only be determined by building and subjecting the design to the actual manufacturing environment. It cannot be exactly calculated. It can only be modeled and approximated mathematically. In design for manufacturability, a sensitivity measure close to manufacturing sensitivity would be most useful.

4.1.5 Three Sensitivity Concepts

There are three sensitivity concepts which are compared and contrasted in this chapter. The first, manufacturing sensitivity, acts as the desired measure for our manufacturing-oriented sensitivity analysis. It is the conceptual measure of how suited a design is to the manufacturing (or unit service lifetime) environment. Manufacturing sensitivity measures the sensitivity of a unit's manufactured-ensemble performance to changes in the design parameters. The other two, classic sensitivity and *statistical sensitivity,* can be mathematically formulated and calculated, and used by the designer to better understand and control various statistical aspects of the unit's performance. For clarity and ease of reference, the three sensitivities are summarized below.

Manufacturing Sensitivity

During manufacture (and service life), the change in unit statistical performance[1] for a change in a given unit design parameter.

Classic ("Single-Point") Sensitivity

A unit's calculated single-point performance[2] change as a function of a unit parameter[3] change.

Statistical Sensitivity

A unit's calculated statistical performance change as a function of a unit density parameter[4] change.

4.2 ILLUSTRATIVE PROBLEMS

To properly distinguish this chapter from other works on sensitivity, we present the following two simple, yet practical, problems.

Problem 4.1—Voltage Divider

The voltage divider shown in Figure 4.2 is the subject of our first example problem. The output voltage, when $V_{in} = 1V$, is given by

$$V_{out} = \frac{R_2}{R_1 + R_2} \tag{4.1}$$

The normalized classic sensitivity of V_{out} with respect to R_1 is given by

$$SN_{R_1}^{V_{out}}(R_1,R_2) = \frac{R_1}{V_{out}} \frac{\partial}{\partial R_1} V_{out} = \left(\frac{R_1}{R_2/(R_1 + R_2)}\right)\left(-\frac{R_2}{(R_1 + R_2)^2}\right) \tag{4.2}$$

---

[1] Statistical performance examples are performance average value, variance, and yield.
[2] Performance examples (for circuits) are gain, bandwidth, and noise figure.
[3] Unit parameter examples (for circuits) are resistance, inductance, and transistor gm.
[4] Density parameter examples are unit parameter average value, variance, and correlation.

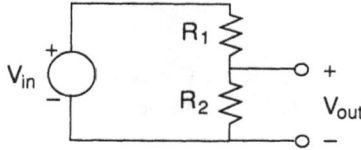

Figure 4.2 Voltage divider used in Problem 4.1.

which when evaluated at $R_1 = R_2$, becomes

$$SN_{R_1}^{V_{out}}(R_1, R_2) = -\frac{1}{4}$$

Likewise, the normalized classic sensitivity of V_{out} with respect to parameter R_2 is

$$SN_{R_2}^{V_{out}}(R_1, R_2) = \left(\frac{R_1}{R_2(R_1 + R_2)}\right)\left(+\frac{R_1}{(R_1 + R_2)^2}\right) \tag{4.3}$$

which when evaluated at $R_1 = R_2$, becomes

$$SN_{R_2}^{V_{out}}(R_2, R_2) = +\frac{1}{4}$$

This analysis says that a 1% change in either R_1 or R_2 will result in a $\pm 0.25\%$ change in V_{out}. Is this good or bad? It really depends on expectations for the unit performance. If each parameter varies $\pm 10\%$, the individual variation in Vout will be about $\pm 2.5\%$ for each component's individual contribution. But if the resisitor values are changing simultaneously, possibly in a correlated way as they can during manufacturing, what kind of V_{out} variation can we expect? Single-point sensitivities cannot address this question. We will come back to this problem, but first we introduce our second example, a mechanical problem.

Problem 4.2—Mechanics of a Lug Nut

This example problem is nonlinear and is an extreme case of a type of problem encountered in circuit and system design. This example is used for simplicity and clarity of presentation, not because it, in itself, is the type of problem that is usually encountered. The reader, however, should note that practical problems can exhibit the type of performance illustrated in this example.

The mechanical system shown in Figure 4.3 is analyzed. We wish to determine the sensitivity of the tightness of the automobile wheel as a function of the tightness

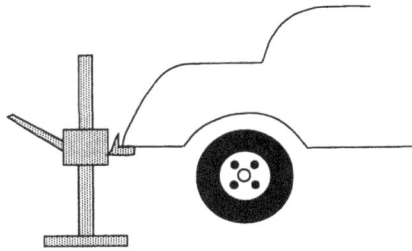

Figure 4.3 The "lug-nut" problem: determining the sensitivity of the tightness of a car wheel as a function of the tightness of the lug nuts holding the wheel.

of the four lug nuts that fasten the wheel to the axle. In the context of the design problems in previous chapters, each lug nut tightness is a unit parameter, and the unit performance is the tightness of the wheel.

To put a mathematical framework around this problem, let p_i ($i = 1,...4$) be the lug nut tightness, such that $0 \le p_i \le 1$. If $p_i = 0$, the lug nut is perfectly tight, and if $p_i = 1$, the lug nut is fully loose. Therefore, the numerical value of p_i indicates the tightness of the lug nut in some sense. Furthermore, let W be a numerical variable indicating the wheel's tightness, such that $0 \le W \le 1$. Again with $W = 0$ indicating perfect tightness and $W = 1$ indicating the wheel is totally loose.

We will model the functional relation between the unit performance, W, and the unit parameters, p_i ($i = 1,...,4$) for this problem as:

$$W(p_1, p_2, p_3, p_4) = \text{minimum}(p_1, p_2, p_3, p_4)$$

The model says that if any one lug nut is tight, the wheel is tight. The wheel is only as loose as the tightest lug nut. A two dimensional plot of W, assuming the wheel has only two lug nuts is shown in Figure 4.4. Although this function is nonlinear in p_1 and p_2, you can see that it is not particularly wild or unusual in its appearance. Practically all circuits and systems are nonlinear in their parameters, even ones that have linear input-output relationships. The behavior of W is not unlike some circuits and systems with parallel or redundant components. W is continuous everywhere, but it is not differentiable everywhere. The properties of W that are exploited in this chapter also apply to differentiable functions.

The performance of this system is a function of the parameter nominal values, $p_{10},...,p_{40}$. Assuming that the nominal values are $P_0 = (0,0,0,0)$ (i.e., all the lug nuts are tightened down) then $W(P_0) = 0$—the wheel is tight. The classic sensitivity of W with respect to each of the p_i evaluated at P_0 is also 0:

$$S_{p_i}^W(0,0,0,0) = 0 \qquad (4.4)$$

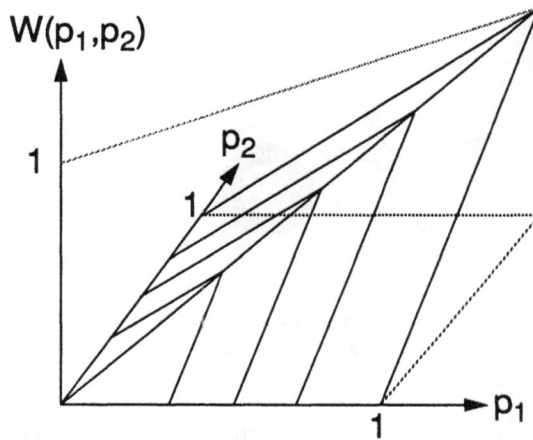

Figure 4.4 The graphical description of $W(p_1,p_2)$ = minimum(p_1,p_2), versus p_1 and p_2.

for all i. From the mechanical analogy this makes sense. The wheel looseness, with all the others lug nuts tight, is not affected by the position of any single lug nut. Mathematically this is correct, but intuitively, this result is disturbing because an analysis using these sensitivities alone might lead us to believe that the wheel tightness is not a function of any of the lug nuts. Hence, this classic sensitivity analysis, isolated from any other reasoning, implies that the lug nuts are not critical parameters to the wheel tightness, which is an incorrect conclusion.

Another important issue in a manufacturability study is the likelihood of a lug nut being loose. For example, if lug nut 1 is put on without a lock washer, it may be more likely that it will be loose during manufacture. Intuitively this will affect the tightness of the wheel, but the classic sensitivity figures cannot incorporate such information into the sensitivity analysis. Essentially, the parameter joint probability density function contains important information that should be considered in the sensitivity analysis. Classic sensitivity studies do not use the parameter probability density function in any way.

4.3 REVIEW OF SENSITIVITY STUDIES

In this section we will review the classic sensitivity definitions, including *large-change* and *multiparameter sensitivities*. We will also show the sensitivity of the voltage divider and the mechanical problem. A statistical-sensitivity measure which overcomes the inherent limitations of traditional sensitivity analysis is presented. It gives the design engineer valuable information regarding the interaction of the design with the manufacturing environment.

4.3.1 Single-Point Sensitivity

The simplest sensitivity, which we call the single-point sensitivity, is the derivative of a differentiable function G(P) (which stands for performance) with respect to a parameter value, p_i, evaluated at some point in parameter space, usually the nominal value. This is the classic unnormalized sensitivity:

$$S_{p_i}^G = \frac{\partial G(P)}{\partial p_i} \tag{4.5}$$

equals the change in G(P) with respect to changes in p_i. Recall that $P = (p_1, p_2,...p_N)$ is the parameter vector, and p_i is the ith parameter. Also $S_{p_i}^{G(P)}$ is a function of the point in parameter space where it is evaluated, and we recognize this with the notation

$$S_{p_i}^{G(P)} = S_{p_i}^{G(P)}(P)$$

where P is the evaluation point. We note that $S_{p_i}^{G(P)}$ can be a strong function of P.

This sensitivity has been used successfully to determine the local behavior of performance with respect to the parameters. But this sensitivity figure determines the performance variation due to only a single parameter.

Voltage Divider

Applying the classic sensitivity to the voltage divider of Problem 4.1 (when $R_1 = R_2$), we see that

$$SN_{R_1}^{V_{out}} = -\frac{1}{2} \qquad SN_{R_2}^{V_{out}} = \frac{1}{2}$$

The information conveyed here is that the performance is equally affected by small changes in either resistor. And the effect is negative for R_1 and positive for R_2. The conclusion is that each needs to be controlled about equally for good performance control.

Lug Nut

Applying the classic sensitivity figure to our lug-nut case, Problem 4.2, illustrates a potential weakness. Applying the definition, we see that $S_{p_i}^W(p_1,p_2,p_3,p_4)$ equals one if p_i is the only minimum of the set of parameters (p_1,p_2,p_3,p_4), and equals zero otherwise. Thus, we see that $S_{p_i}^W$ is almost always zero. If the nominal value of P is (0,0,0,0), $S_{pi}^{W}(0,0,0,0) = 0$ for all i. This is misleading in that it gives the impression that W is flat around the point (0,0,0,0). Looking at the plot of W in Figure 4.4, this is not the case—it is flat only along the axes.

The problem is that classic sensitivity records the performance changes for one parameter at a time and for small variations in each parameter. Although for many types of performance this is adequate (as in the voltage-divider example), this sensitivity does not capture the manufacturing environment, where all parameters are varying simultaneously. This mechanical example with P = (0,0,0,0) is particularly fussy about how the parameters are changed. Several variations on the classic sensitivity have been developed to overcome these problems.

4.3.2 Multiparameter Sensitivity

Because performance is usually a function of many variables, the multiparameter sensitivity was developed to capture the sensitivity effects due to the simultaneous changes in the parameters. The multiparameter sensitivity of $G(p_1, p_2, ... p_N)$ is expressed mathematically by the total differential

$$\text{M.P.S}_{pi}^{G} = \sum_{i=1}^{N} \left(\frac{\partial G}{\partial p_i} \right) dp_i$$

Sometimes this is normalized by $G(P_0)$, but for our purposes this total differential is satisfactory. The multiparameter sensitivity is also a function of the point in parameter space on which it is evaluated, shown by

$$\text{M.P.S}_{pi}^{G} = \text{M.P.S}_{pi}^{G}(P)$$

This sensitivity can be written in terms of the single-point sensitivities as

$$\text{M.P.S}_{pi}^{G}(P) = \sum_{i=1}^{N} S_{pi}^{G}(P) dp_i$$

Voltage Divider

Applying the normalized multiparameter sensitivity to the voltage divider (when $R_1 = R_2$) gives

$$\text{M.P.S}_{R}^{V_{out}} = -\frac{1}{4} + \frac{1}{4} = 0$$

This indicates that, on the average, the variations in R_1 and R_2 will cancel. This, of course, assumes that R_1 and R_2 are statistically independent, which is not always the case. This sensitivity figure does show the effects of cancelling that might statistically occur during manufacturing.

Lug Nut

Applying the multiparameter sensitivity to our lug nut problem we can determine that $M.P.S_P^W(p_1,p_2,p_3,p_4) = 0$ if there is an i and j such that $p_i = p_j$ and both are less than or equal to all the other p_i, and $M.P.S_P^W(p_1,p_2,p_3,p_4) = 1$ if there is an i such that p_i is less than all the other p. Specifically, the multiparameter sensitivity of W evaluated at $P = (0,0,0,0)$ is

$$M.P.S_P^W(0,0,0,0) = 0$$

From this result we might incorrectly conclude that even with the simultaneous changes of the parameters, the function W is insensitive to all the parameters. However this sensitivity still only entails an examination of performance along the parameter axes.

4.3.3 Large-Change Sensitivity

The large-change sensitivity tracks the variation of performance as a parameter is subjected to variations that need not be small. Define a perturbed parameter vector as

$$\Delta_i P = (p_1,p_2,...p_i + \Delta_i,...p_N)$$

then the large-change sensitivity is

$$L.C.S_{P_i}^G(P,\Delta_i P) = G(\Delta_i P) - G(P)$$

This sensitivity can be useful because it does involve large changes in parameters like might be encountered during manufacture, rather than the infinitesimal changes used in the other sensitivities described.

Voltage Divider

Assuming that each parameter is perturbed by $+5\%$, $R_1 = R_2$, and $V_{in} = 1.0V$, the sensitivities are given by

$$L.C.S_{R_1}^{V_{out}}(R_1,R_1) \approx -0.01$$

$$L.C.S_{R_2}^{V_{out}}(R_2,R_2) \approx 0.01$$

These show that the output voltage swing will be about 1% in magnitude with a 5% change in any individual parameter. This is different than will be predicted by the classic sensitivity, which is

$$\frac{0.0025\dfrac{\Delta V_{out}}{V_{out}}}{\%\Delta R}(5\%\,\Delta R)(0.5 V_{out}) \;=\; 0.00625\Delta V_{out}$$

Differences between the classic sensitivity and the large change sensitivity demonstrate the need for a sensitivity figure to examine the performance space more thoroughly. For a manufacturing study, the large-change sensitivity will more closely approximate the kind of parametric changes that occur during manufacture.

Lug Nut

Applying the large-change sensitivities to the lug nut problem, we obtain an interesting result:

$$\text{L.C.S}_{P_i}^W(P) \;=\; S_{P_i}^W(P)$$

For this problem the infinitesimal single-point sensitivity and the large-change sensitivity are the same.

4.3.4 Multiparameter Large-Change Sensitivity

Our final sensitivity considers the simultaneous large change of all the parameters to determine the manufacturing sensitivity. It is similar to the multiparameter sensitivity just presented. Briefly, this sensitivity is defined as

$$\text{M.P.L.C.S}_{P_i}^G \;=\; \sum_{i=1}^{N} \text{L.C.S}_{P_i}^G(P,\Delta_i P)$$

Voltage Divider

Applying the large-change multiparameter classical sensitivity to the voltage divider, we see that the sensitivity is approximately zero. This again tells the designer the symmetry in this problem and also that the effects of the large-change variations will, on the average, cancel out. It is assumed that the parameters are independently and identically distributed. If this is not true, the present analysis will not give quantitative results about the design performance sensitivity.

Lug Nut

Applying this sensitivity method to our lug-nut problem we see that

$$M.P.L.C.S_{P_i}^W(0,0,0,0) = 0$$

Once again, the restriction to the parameter axes made by these sensitivity measures is a weakness illustrated by this example. In many problems, like the voltage divider, they can give useful information to the designer. However, by their nature they do not account for the full span of parameter variations that will occur during manufacture; that is, they do not fully explore the performance space of the unit, nor do they account for the statistical description of the variables.

The statistical sensitivity, which will be introduced in Section 4.4, addresses the difficulties presented in both the voltage-divider and mechanical example problems. Before introducing the statistical sensitivity, a sensitivity-reduction process that has been developed and popularized by G. Taguchi [5] and others is presented.

4.3.5 Performance Variance Reduction, Taguchi Methods [5]

The methods commonly called Taguchi Methods were pioneered by Dr. Taguchi in the 1950s and early 1960s. From our point of view, these methods accomplish one important aspect of sensitivity analysis that is missing when using the classical sensitivity figures mentioned above. The Taguchi Methods require a systematic search of the performance space at parameter points off the parameter space axes. This basic innovation is a great asset to the Taguchi Methods. Also, these methods attempt to minimize, by using statistical experiment design, the number of performance evaluations needed to determine and minimize performance sensitivity.

The sensitivity optimization used in the Taguchi Method assumes that the unit parameters can be divided into two groups. The first group is the control parameters (factors). These parameters are ones that inherently affect the performance variation and determine, not the absolute unit performance, but the performance variance.

The second group is the adjustment parameters. These parameters are ones that mostly affect the performance average value. While splitting of parameters into these two groups is sometimes not possible, this has been accomplished in many published examples such as integrated-circuit fabrication [6], wave soldering [7], manufacturing processes [8], computer systems [9], and many others [10].

After the parameters have been decomposed into the control and adjustment groups, the Taguchi Method proposes a two-step procedure of sensitivity analysis and reduction. Step 1 calls for determining the control parameter values that maximize the so-called signal-to-noise ratio, which is the normalized ratio of the performance mean value to the performance standard deviation. Step 2 involves determining the adjustment parameter values which move the performance average value to meet the desired specification.

Selection of the appropriate factors and finding their optimal values is accomplished using orthogonal array experiments. First, the parameters are identified. For instance, assume we have four parameters: p_1, p_2, p_3, p_4. Levels are assigned to each parameter. These are essentially large changes assigned to each parameter. For our problem (and following the form of the lug-nut problem), we will assign levels to each parameter of p_{i0}, $p_{i0} + \Delta_i$, and $p_{i0} + 2\Delta_i$. Thus, we are assuming only positive variation of each parameter for this example, although in general the variation can also be $p_{i0} - \Delta_i$.

Next, an orthogonal-matrix experiment is formulated for these parameters and their levels. The details of this formulation are beyond our intended scope, and the reader is referred to any book on experiment design for the details [2, 5, 10], for example. The *experiment matrix* has columns for each of the parameters, and rows for each experiment that is to be simulated. The orthogonality of the experiment occurs because every column of the matrix is different, and for every pair of columns, all combinations of factor levels occur an equal number of times. Table 4.1 shows a possible orthogonal-matrix experiment for our mechanical problem.

This experimental matrix then represents nine sets of parameter values for which the performance will be evaluated. A graph of these performance points for parameters 1 and 2 is shown in Figure 4.5. Looking at the experiment matrix or at Figure 4.5, we see that this method requires the evaluation of unit performance at points other than on the parameter axes.

Next, the signal-to-noise ratios are calculated at each of the experiment points. The signal-to-noise ratios (S/N) are essentially the inverse of the normalized

Table 4.1
An Orthogonal Matrix Experiment for the Lug-Nut Problem

Experiment Number	Parameter 1	Parameter 2	Parameter 3	Parameter 4
1	p_{10}	p_{20}	p_{30}	p_{40}
2	p_{10}	$p_{20} + \Delta_2$	$p_{30} + \Delta_3$	$p_{40} + \Delta_4$
3	p_{10}	$p_{20} + 2\Delta_2$	$p_{30} + 2\Delta_3$	$p_{40} + 2\Delta_4$
4	$p_{10} + \Delta_1$	p_{20}	$p_{30} + \Delta_3$	$p_{40} + 2\Delta_4$
5	$p_{10} + \Delta_1$	$p_{20} + \Delta_2$	$p_{30} + 2\Delta_3$	p_{40}
6	$p_{10} + \Delta_1$	$p_{20} + 2\Delta_2$	p_{30}	$p_{40} + \Delta_4$
7	$p_{10} + 2\Delta_1$	p_{20}	$p_{30} + 2\Delta_3$	$p_{40} + \Delta_4$
8	$p_{10} + 2\Delta_1$	$p_{20} + \Delta_2$	p_{30}	$p_{40} + 2\Delta_4$
9	$p_{10} + 2\Delta_1$	$p_{20} + 2\Delta_2$	$p_{30} + \Delta_3$	p_{40}

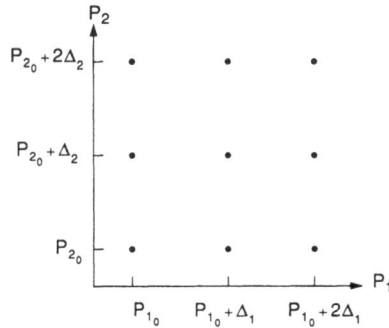

Figure 4.5 The experimental points for two parameters as set by the orthogonal experiment matrix.

multiparameter large-change sensitivities evaluated at each of the experiment points. Other sensitivity measures can be used, but they must be derived from the data taken in the orthogonal-matrix experiment. The optimization principle proposed by Taguchi is then:

1. To find the parameter vector in the experiment which maximizes S/N, or usually equivalently, the parameter vector in the experiment which minimizes the large change multiparameter sensitivities;
2. To modify the mean performance using the adjustment parameters, also identified in the matrix experiment, to achieve the desired performance.

There are many variations on this theme; however, this presentation should give adequate insight.

Lug Nut

Apply the Taguchi Method to the lug-nut problem. Recall that the performance W is given by

$$W(p_1, p_2, p_3, p_4) = \text{minimum}(p_1, p_2, p_3, p_4)$$

For this case, we assigned levels of 0.1 and 0.2 to each variable. The process of assigning levels is usually not well described; however, the levels assigned generally represent the designer's best guess of possible variations that each parameter will encounter during manufacture or during the unit's lifetime. We choose a nominal set of parameters. The results for this problem will change for different choices of P_0, but to be consistent with the other uses of this example, we choose $P_0 = (0,0,0,0)$. Table 4.2 shows the orthogonal parameter matrix and the performance evaluation.

Table 4.2
Orthogonal Parameter Matrix and Performance Evaluation

Experiment Number	Parameter 1	Parameter 2	Parameter 3	Parameter 4	Performance W
1	0.0	0.0	0.0	0.0	0.0
2	0.0	0.1	0.1	0.1	0.0
3	0.0	0.2	0.2	0.2	0.0
4	0.1	0.0	0.1	0.2	0.0
5	0.1	0.1	0.2	0.0	0.0
6	0.1	0.2	0.0	0.1	0.0
7	0.2	0.0	0.2	0.1	0.0
8	0.2	0.1	0.0	0.2	0.0
9	0.2	0.2	0.1	0.0	0.0

Since this matrix experiment has at least one of the four variables set to its nominal value, the performance (wheel tightness) is always zero (i.e., the wheel is always tight). We might conclude from this set of experiments that the parameters do not affect performance.

For this rather tough example problem, these initial experiment-design methods fail to illustrate to the designer the relationship that performance has to the parameters. Their strength is the attempt to interrogate the performance space with large-change parameter variations that are not confined to the axes. And, in general, this is a great improvement from the classical sensitivity measures because manufacturing variations in parameters are certainly not constrained to single variations on the parameter axis. However, the parameter variation is constrained by the experiment design and does not in any way reflect the statistical variations that can occur during manufacture.

A good practitioner of the experiment-design approach will not be satisfied with these nine experiments if all the experimental outcomes are the same. Basically, the number of degrees of freedom in the performance has been under-estimated, and the obvious thing to do is to use a more complex matrix experiment. This will result in an experiment conducted with no parameter value at the nominal value; hence, the performance will be nonzero. However, we will not pursue this problem further here.

The experiment-design approach to performance evaluation is best applied in situations where an experiment (or a simulation in our case) is very costly or time-consuming. Then the orthogonal matrix of experiments determined by experiment design is probably the best choice for the small number of experiments. For most types of well-behaved performances, and for simple statistical relations among the parameters, the information gathered from these small number of experiments is probably as good as can be done. However, our example shows that these efficient experimental strategies often do not give the designer all of the information that is needed.

The fundamental weakness of these methods, independent of the number of degrees of freedom in the problem or the size of the matrix experiment, is that these methods do not take into account the parameter joint probability density in any systematic way. The levels are chosen by the designer to represent the variation of each parameter. But the levels are applied to the experiment in a random and uncorrelated manner. The basic underlying assumption for the application of levels is that the parameters are each uniformly distributed and jointly independent. This can be a poor assumption for certain circuit and system problems.

4.4 MANUFACTURING SENSITIVITY

There are two essential properties that should be included in a unit's sensitivity study if manufacturing sensitivity is to be determined:

1. The sensitivity needs to include the sense of simultaneous large changes in the parameter values, like those encountered during manufacture or the unit lifetime.
2. The sensitivity needs to include the actual variations encountered during manufacture, including the higher order statistical properties; that is, the parameter variation needs to follow the parameter joint probability density function. For instance if the performance is sensitive to a set of parameters that are impossible to encounter during the unit's manufacture or lifetime, then this sensitivity should be of no interest to the designer. Certainly an impossible-to-encounter sensitivity need not be considered or manipulated in the design.

These issues are included in what we call the *statistical sensitivity* of a unit. The performance statistical sensitivity and the *yield statistical sensitivity* (or yield sensitivity) are described in the next two sections.

4.4.1 Performance Statistical Sensitivity

A key to the philosophy of the statistical sensitivities is to consider a parameter vector P_0, not as an isolated point in parameter space, as is the case when we evaluate the performance, $G(P_0)$, but to consider P_0 as the nominal or average value of the parameter joint PDF. We can divide an arbitrary parameter P into two parts, its mean and its variation about the mean:

$$\mathcal{P} = \Delta\mathcal{P} + P_0$$

where $P_0 = (p_{10}, p_{20}, \ldots p_{n0}) = E(\mathcal{P})$, and

$$\Delta\mathcal{P} = \mathcal{P} - P_0$$

E($\mathcal{P}$) is the mean or average value of $\mathcal{P}$, and E(.) is the expectation operation. For this case

$$f_{\mathcal{P}}(P) = f_{\Delta\mathcal{P}}(P) * \delta(P - P_0)$$

where $f_{\Delta\mathcal{P}}(P)$ is the zero-mean density, $\delta(P - P_0)$ is the n-dimensional Dirac delta function, and * represents n-dimension convolution. The mathematics convey that the parameter density $f_{\mathcal{P}}(P)$ is equivalent to a density centered about zero (i.e., zero mean) displaced to the parameter vector mean value, P_0.

Statistical design usually entails the selection of P_0—not $\mathcal{P}$. The manufacturing environment usually sets $f_{\Delta\mathcal{P}}(P)$. Therefore by choosing P_0, and knowing $f_{\Delta\mathcal{P}}(P)$ we know the parameter statistics for $\mathcal{P}$. It is a mistake to think that the designer can directly specify $\mathcal{P}$ as a design variable. In our manufacturing model, $\mathcal{P}$ is a random variable and as such cannot be set by the designer. However P_0, the mean value of the parameters, is not random, and hence can be used as a design variable.

There is another concept to address. The performance G($\mathcal{P}$), which is the performance of a given manufactured unit, is also a random variable, because $\mathcal{P}$ is random during manufacture. The nominal performance $G_0 = G(P_0)$ is calculable but is of little value in determining the statistics of G($\mathcal{P}$). In fact, because G is a nonlinear function,

$$E(G(\mathcal{P})) \neq G(E(\mathcal{P})) = G(P_0)$$

Thus, the nominal performance is not even the same as the average performance of the manufactured unit. In order to characterize the manufactured performance, G($\mathcal{P}$), we must deal with some ensemble measure of its statistics, such as its average, its variance, or the yield. The concept that manufacturing performance is a random variable, or vector, is of fundamental importance in characterizing a unit's manufacturing performance. The only way to characterize a random variable is with its statistics. Stating that a random variable can assume a given value is of little use, the real question is how often does it assume the value, or more useful, what is the probability it assumes values in a given value range. This concept should also be applied to random performance.

The statistical sensitivities are concerned with some statistical measure of the unit performance as a function of the parameter average values, given $f_{\Delta\mathcal{P}}(P)$. For the performance sensitivity, we consider the statistical average of the performance. Remember that G($\mathcal{P}$), the unit performance, is defined as

$$\text{Performance} = G(\mathcal{P}) = G(\Delta\mathcal{P} + P_0)$$

Then the unit average performance, called $\bar{G}(\mathcal{P})$, is given by

$$\bar{G}(\mathcal{P}) = \int_{-\infty}^{\infty} G(\Delta\mathcal{P} + P_0)f_{\mathcal{P}}(\Delta\mathcal{P} + P_0)d(\Delta\mathcal{P} + P_0)$$

$$= \bar{G}(P_0)$$

substituting in for $f_{\mathcal{P}}(P)$ we obtain

$$\bar{G}(P_0) = \int_{-\infty}^{\infty} G(\Delta\mathcal{P} + P_0)[f_{\Delta\mathcal{P}}(\Delta\mathcal{P}) * \delta(\Delta\mathcal{P} - P_0)]d\Delta\mathcal{P}$$

The average performance is clearly a function of P_0. We notationally recognize this functionality by writing $\bar{G}$ as $\bar{G}(P_0)$.

Using the Fundamental Theorem of Monte Carlo, we obtain an unbiased estimate of the unit average performance as $\hat{G}$:

$$\hat{G}(P_0) = \frac{1}{M}\sum_{i=1}^{M} G(P_i)$$

where the P_i are samples chosen according to $f_{\mathcal{P}}(P)$.

Next, define the statistical sensitivity of the average performance with respect to a parameter in P_0 as

$$S_{P_{i0}}^{\bar{G}(P_0)} = \frac{\partial}{\partial p_{i0}}\bar{G}(P_0)$$

This sensitivity figure has an inherent advantage because it requires a statistically complete exploration of the performance space, as specified by the parameter joint PDF $f_{\mathcal{P}}(P)$. This addresses the two manufacturing-oriented limitations of classic sensitivity figures: it considers the large change of variables, and it includes the parameter joint PDF.

A comparison of the single-point sensitivity of performance with the statistical sensitivity of average performance provides a better understanding of these sensitivity measures [4]. Assume for simplicity of notation that there are only two parameters ($n = 2$). We will use normalized variables such that the nominal values are all zero and their distributions are independent and uniformly distributed with variation $\pm 1/2$. (This simplification is not necessary but is used here to simplify the notation.) Also, assume that performance can be written in a series expansion normalized about the nominal performance, G_0, as

$$G(p_1, p_2) = G_0 + a_1 p_1 + a_2 p_2 + a_{12} p_1 p_2 + a_{11} p_1^2$$
$$+ a_{22} p_2^2 + a_{111} p_1^3 + a_{122} p_1 p_2^2$$
$$+ a_{112} p_1^2 p_2 + \dots$$

The partial derivative of G with respect to p_1 is included in the classical sensitivity S_{pi}^G. It is expanded as

$$\frac{\partial G(p_1, p_2)}{\partial p_1} = 0 + a_1 + 0 + a_{12} p_2 + 2a_{11} p_1$$
$$+ 0 + 3a_{111} p_1^2 + a_{122} p_2^2$$
$$+ 2a_{112} p_1 p_2 + \dots$$

If this derivative is evaluated at the nominal values, $p_1 = 0$ and $p_2 = 0$, which is usually the case, the result is

$$\frac{\partial G(p_1, p_2)}{\partial p_1} = a_1$$

Now the statistical performance sensitivity of average performance can be derived. For the statistical interpretation, each parameter is divided into its mean and its zero-mean variation (i.e., $p_1 = \Delta p_1 + p_{1_0}$, and $p_2 = \Delta p_2 + p_{2_0}$, and therefore,

$$\frac{\partial \bar{G}(p_{1_0}, p_{2_0})}{\partial p_{1_0}} = \frac{\partial}{\partial p_{1_0}} \left(\int_{-0.5}^{0.5} \int_{-0.5}^{0.5} G(p_1, p_2) f(p_1, p_2) dp_1 dp_2 \right.$$

$$= \frac{\partial}{\partial p_{1_0}} \left(\int_{-0.5}^{0.5} \int_{-0.5}^{0.5} (G_0 + a_1 p_1 + a_2 p_2 + a_{12} p_1 p_2 + a_{11} p_1^2 \right.$$

$$+ a_{22} p_2^2 + a_{111} p_1^3 + a_{122} p_1 p_2^2$$

$$+ a_{112} p_1^2 p_2 + \dots) dp_1 dp_2$$

Taking the derivative inside the integral, noting that $dp_1 = d(p_{1_0} + \Delta p_1)$, and evaluating at the nominal parameter value, $p_{1_0} = p_{2_0} = 0$,

$$\frac{\partial \bar{G}(p_1, p_2)}{\partial p_{1_0}} = \int_{-0.5}^{0.5} \int_{-0.5}^{0.5} (0 + a_1 + 0 + a_{12} \Delta p_2 + 2a_{11} \Delta p_1$$

$$+ 0 + 3a_{111} \Delta p_1^2 + a_{122} \Delta p_2^2$$

$$+ 2a_{112} \Delta p_1 \Delta p_2 + \dots) d\Delta p_1 d\Delta p_2$$

and since $E(\Delta p_1) = E(\Delta p_2) = 0$, where $E(.)$ is the expectation operator,

$$\frac{\partial \bar{G}(p_1,p_2)}{\partial p_{10}} = a_1 + 3a_{11}E(\Delta p_1^2) + a_{122}E(\Delta p_2^2)$$

$$+ 2a_{112}E(\Delta p_1 \Delta p_2) + \ldots$$

The higher order terms in the series expansion of performance are maintained in the statistical average. The parameters p_1 and p_2 in the performance derivative are replaced by their appropriate moments. Thus, the higher order behavior of the derivative is preserved. This is in contrast to the classical sensitivity where the parameters p_1 and p_2 in the performance are replaced by the nominal values.

This derivation shows that performance sensitivity and the average performance statistical sensitivity are not the same numerically. In fact, they can be distinctly different—even having opposite signs. Hence, the single-point optimized design and the statistically optimized design can lead to unlike solutions, as many examples in this book illustrate.

The average performance statistical sensitivity and the performance sensitivity are the same only when there is no parameter variation (i.e., when $f_{\mathscr{P}}(P,P_0) = \delta(P - P_0)$). Classical sensitivities are simply a special case of statistical sensitivities.

Lug Nut

We will now apply the notion of average performance statistical sensitivity to our lug-nut problem. To simplify the notation and the calculations, assume that there are two parameters instead of the usual four. In order to perform a statistical analysis, we need to know the parameter statistical model. For this example assume that the parameters are independent and uniformly distributed over the interval $[0,.1]$. The parameter joint density

$$f_{\Delta\mathscr{P}}(\Delta p_1,\Delta p_2) = \begin{cases} 100 \text{ if } 0 \leq \Delta p_1 \leq 0.1 \text{ and } 0 \leq \Delta p_2 \leq .1 \\ 0 \text{ otherwise} \end{cases}$$

The parameters are modeled as $p_1 = p_{10} + \Delta p_1$ and $p_2 = p_{20} + \Delta p_2$. Therefore, the parameter joint density is

$$f_{\mathscr{P}}(p_{10} + \Delta p_1, p_{20} + \Delta p_2)$$
$$= \begin{cases} 100 \text{ if } p_{10} \leq p_{10} + \Delta p_1 \leq p_{10} + 0.1 \text{ and } p_{20} \leq p_{20} + \Delta p_2 \leq p_{20} + 0.1 \\ 0 \text{ otherwise} \end{cases}$$

$W(p_1,p_2) = \text{minimum}(p_1,p_2)$ is the performance function. The average performance, $\bar{W}$, is given by

$$\bar{W}(p_{10},p_{20}) = \int_{-\infty}^{\infty}\int_{-\infty}^{\infty} \min(p_{10} + \Delta p_1, p_{20} + \Delta p_2) f_{\Delta \mathcal{P}}(\Delta p_1, \Delta p_2) d\Delta p_1 d\Delta p_2$$

$$\int_0^{0.1}\int_0^{0.1} \min(p_{10} + \Delta p_1, p_{20} + \Delta p_2)(100) d\Delta p_1 d\Delta p_2$$

This expression has different solutions depending on the relation between p_{10} and p_{20}. If $p_{10} = p_{20}$, we get

$$\bar{W}(p_{10},p_{20}) = 100 \int_0^{0.1} \left[\left(\int_{p_{10}}^{p_{10}+\Delta p_2} (p_{10} + \Delta p_1) d\Delta p_1 + \int_{p_{10}+\Delta p_2}^{p_{10}+0.1} (p_{20} + \Delta p_2) d\Delta p_1 \right) \right] d\Delta p_2$$

Upon evaluation, $\bar{W}(p_{10},p_{20}) = p_{10} + 0.2 / 6$. The dependence of the average performance on the density parameters, $p_{10} = p_{20}$, is explicitly shown. Also, at $p_{10} = p_{20} = 0$, the average performance (the average wheel tightness) is $0.2 / 6$ and not zero as has been observed in all the single-point analyses of this problem thus far. This measure, $\bar{W}$, involves the weighted average of the points in performance space, and the weighting is given by $f(p_{10},p_{20})$. (The average wheel tightness will be influenced by the presence or absence of lock washers on the lug nuts.) This is an intuitively correct result. It is important to use a performance measure, like $\bar{W}$, which has this property when investigating manufacturing sensitivity.

We can now determine the average performance statistical sensitivity for this problem when $p_{10} = p_{20}$, as

$$S_{p_{10}}^{\bar{W}} = \frac{\partial}{\partial p_{10}}(\bar{W}(p_{10},p_{20})) = 1.0$$

Thus, the statistical sensitivity reveals that average performance is sensitive to the parameter p_{10} (and p_{20} since $p_{10} = p_{20}$). Also, the average performance is strongly influenced by these density parameters as shown by the numerical value of 1.0. The sign indicates that as the parameter becomes larger, the average performance (wheel looseness) also increases.

In summary, we see that the statistical sensitivity exhibits the two important characteristics of a sensitivity figure for manufacturing (see also Exercise 4.1):

1. It includes large changes in the parameter values.
2. It includes the combinatorial aspects of the parameters due to the joint PDF.

4.4.2 Yield Statistical Sensitivity

As noted in Chapter 3, Yield (Y) is defined as

$$Y = \int_{-\infty}^{\infty} \text{Accept}(P) f_{\mathcal{P}}(P) dP$$

where $P = (p_1, p_2, \ldots p_n)$ is the parameter vector, $f_{\mathscr{P}}(P)$ is the parameter joint PDF, and accept(P) is the acceptance function in parameter space. According to Section 4.4.1, we can divide $\mathscr{P}$ into its constant mean P_0 and its zero-mean variation $\Delta \mathscr{P}$, as

$$\mathscr{P} = P_0 + \Delta \mathscr{P}$$

Since $d\mathscr{P} = d(P_0 + \Delta \mathscr{P}) = 0 + d\Delta \mathscr{P} = d\Delta \mathscr{P}$, we can write yield explicitly as a function of P_0 as

$$Y(P_0) = \int_{-\infty}^{\infty} \text{Accept}((P_0 + \Delta P)f_{\mathscr{P}}(P_0 + \Delta P))d\Delta P$$

Further, we can define the statistical yield sensitivity, or the yield sensitivity, as

$$S_{p_{i_0}}^Y = \frac{\partial}{\partial p_{i_0}} Y(P_0) = \frac{\partial}{\partial p_{1_0}} \int_{-\infty}^{\infty} \text{Accept}((P_0 + \Delta P)f_{\mathscr{P}}(P_0 + \Delta P))d\Delta P$$

This expression looks somewhat complicated, and if there were not a way to approximately calculate it, its use would be limited. An efficient method for calculating this yield sensitivity is given in Section 4.6. This sensitivity is truly statistical in nature since yield is a function of the parameter density function. Also, yield is a weighted average of all the performances encountered during manufacture (as compared with the unit specification) and weighted by the parameter joint density function.

The interpretation of the yield sensitivity is straightforward. If the yield sensitivity is small, then the yield is not a strong function of the nominal value of that parameter. This kind of information is very useful to the designer because it alerts the designer to the brinksmanship type of design described in Chapter 1. Yield sensitivity and its reduction should be a key step in the design of manufacturable units.

Other sensitivities besides parametric sensitivities can be tested with yield sensitivity. The accept function is actually dependent on many variables other than the parameter vector, P. For instance,

$$\text{Accept}(P) = \text{Accept}(P, \text{Str}, \text{Spec}, \ldots)$$

where P is the parameter vector, Str is the unit structure, and Spec is the specifications. There are statistical parameters that affect yield other than the parameter nominal values. For instance,

$$f_{\mathscr{P}}(P) = f_{\mathscr{P}}(P, P_0, T, \sigma, \rho, \ldots)$$

where P_0 is the parameter nominal values $E(P)$, T is the parameter tolerances, σ is the parameter variance, and ρ is parameter correlation coefficients.

Yield is more completely written as

$$Y = \int_{-\infty}^{\infty} \text{Accept}(P,\text{Str},\text{Spec})f_{\mathscr{P}}(P,P_0,T,\sigma,\rho,...)dP$$

Therefore, Yield = Y = Y(Str,Spec,P_0, T, σ, ρ,...) is a function of many parameters, both statistical parameters (skewness, kurtosis, number of kernels in the KDE model) (see Chapter 6) and unit "parameters" like structure.

Three possible sensitivity figures are:

$$S_{\text{Spec}}^{Y} = \frac{\partial}{\partial \text{Spec}}Y$$

$$S_{T}^{Y} = \frac{\partial}{\partial T}Y$$

$$S_{\text{Str}}^{Y} = \frac{\partial}{\partial \text{Str}}Y$$

assuming the derivatives exist.

If we write out one of these expressions in detail, we get

$$\frac{\partial Y}{\partial T} = \frac{\partial}{\partial T}\int_{-\infty}^{\infty} \text{Accept}(P,\text{Str},\text{Spec})f_{\mathscr{P}}(P,P_0,T,\sigma)dP$$

$$= \int_{-\infty}^{\infty} \text{Accept}(P,\text{Str},\text{Spec})\frac{\partial}{\partial T}f_{\mathscr{P}}(P,P_0,T,\sigma)dP$$

Since $f_{\mathscr{P}}(P)$ may be discontinuous with respect to tolerance, the derivatives may not analytically exist at some points. We may get around this by defining a large change derivative as $\Delta f_{\mathscr{P}}(\Delta T) = f_{\mathscr{P}}(T) - f_{\mathscr{P}}(T + \Delta T)$. Plugging this back into the yield equation, we get a large-change yield sensitivity.

4.5 PERFORMANCE VARIANCE SENSITIVITY

This work is primarily focused upon unit yield and its improvement. Yield is a good measure of the ability of a unit to perform well in the manufacturing environment. However, yield does not give total information about the performance. Figure 4.6 shows two possible distributions of performance for a unit [2].

Distribution A shows performance that is peaked around the target value. Distribution B shows performance that is almost uniformly distributed within the acceptable performance interval. From the lower and upper performance specifications as given, each performance distribution gives the same yield. But are these two performance distributions equally desirable? Performance distribution A is

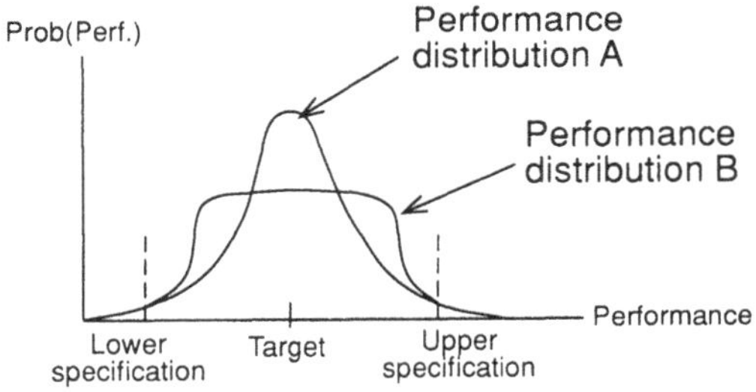

Figure 4.6 Two possible distributions of performance for a manufactured unit.

probably better, because the performance variance is less. A unit that exhibits the performance distribution A might be considered to have more "quality" than a unit with performance distribution B [2]. Quality measurement and improvement is an issue in unit design that is presently being addressed in many arenas. Quality is difficult to define, but it involves reliability, robustness, dependability, and performance variance from unit to unit, and over time. Therefore, from a quality viewpoint, there is a motive to measure and minimize performance variance during manufacture.

Unit performance variance is given as

$$\sigma^2 = \int_{-\infty}^{\infty} \ldots \int_{-\infty}^{\infty} (G(P) - G_0)^2 f_{\mathcal{P}}(P) dP$$

where

$$G_0 = E(P) = \int_{-\infty}^{\infty} \ldots \int_{-\infty}^{\infty} G(P) f_{\mathcal{P}}(P) dP \qquad (4.6)$$

and $G(P)$ stands for the performance of the unit with parameters P. As stated before, the unit variance is a function of the unit parameter density, $f_{\mathcal{P}}(P)$, where $\mathcal{P} = P_0 + \Delta \mathcal{P}$, and $P_0 = E(\mathcal{P})$ is the constant part of the parameter vector. Therefore, it is more appropriate to write the performance variance as

$$\sigma^2(P_0) = \int_{-\infty}^{\infty} \ldots \int_{-\infty}^{\infty} (G(P_0 + \Delta \mathcal{P}) - G_0)^2 f_{\mathcal{P}}(P_0 + \Delta \mathcal{P}) d\Delta \mathcal{P} \qquad (4.7)$$

4.6 PERFORMANCE VARIANCE FACTOR

It will be useful to look at performance variance as a function of the design parameters, P_0. One way to accomplish this is with a type of histogram we have developed for yield analysis. We define a variance factor as

$$\sigma^2(P_0, p_{i0}) = \sigma^2(P_0) \text{ with } f_{\mathscr{P}}(P) = f_{\mathscr{P}}(P) * \delta(p_i - p_{i0}) \tag{4.8}$$

This indicates that the variance factor is the performance variance of the unit with all parameters varying according to the parameter joint distribution, except the ith parameter which is fixed at p_{i0}. An approximate plot of the variance factor versus the parameter value p_{i0}, for a given P_0, is called a variance factor histogram (VFH; an example is given at the end of Section 4.6). By looking at the VFH, the designer can determine values of the individual parameters in P_0, (i.e., p_{i0}, which minimize the performance variance). These concepts can also be used in optimization.

Define the performance variance sensitivity as

$$S_{p_{i0}}^{\sigma^2} = \frac{\partial}{\partial p_{i0}} \sigma^2(P_0)$$

This equation is useful in determining the change in performance variance as a function of the design parameters, p_{i0}. Calculation of this sensitivity is similar to calculating the yield sensitivity described next.

4.7 STATISTICAL SENSITIVITY CALCULATION

The statistical sensitivities, as they are presented, give insight into the statistical performance or yield of a unit as a function of the density parameters used to define the manufacturing environment. But if they cannot be efficiently computed, they will be of little use to the design engineer. In this section we address the issue of calculating these sensitivities and present a method which efficiently calculates an approximation to the sensitivities.

Define a large-change average performance sensitivity as

$$\Delta \bar{G}(p_i, \Delta p_i) = \bar{G}(p_i) - \bar{G}(p_i + \Delta p_i)$$

This is fine conceptually, but from a simulation point of view, it is very costly. An unbiased estimate to $\bar{G}$ can be obtained from a Monte Carlo simulation as

$$\hat{G} = \frac{1}{M} \sum_{i=1}^{M} G(P_i)$$

where the P_i are samples of P taken according to $f_{\mathscr{P}}(P)$.

It will take M evaluations of $G(P_i)$ to get $\hat{G}$. To calculate the large-change sensitivity, this must be done twice, thus taking 2M calculations. This has to be done for each parameter. Thus, if a sensitivity is required for all parameters, this will take $(2 \times M \times n)$ performance evaluations. Depending on the confidence needed in the estimate, M might be 500, and if there are 10 parameters, this would require 10,000 evaluations of performance. Although this in itself is not always prohibitive, this number of evaluations for a complex performance may be impractical. There is an alternative to this approach that can calculate an approximation to the large-change statistical sensitivities for all parameters, in just M performance evaluations.

4.7.1 Performance and Yield Factor

The key to calculating the statistical sensitivities is the average performance factor and the yield factor. They are given as

$$\bar{Y}(p_{i0}) = \int_{-\infty}^{\infty} ... \int_{-\infty}^{\infty} \text{accept}(P) f_{\mathscr{P}}(p_1,...,p_{i-1},p_{i0},p_{i+1},...,p_n)$$
$$dp_1...dp_{i-1}dp_{i+1}...dp_n$$

$$\bar{G}(p_{i0}) = \int_{-\infty}^{\infty} ... \int_{-\infty}^{\infty} \bar{G}(P) f_{\mathscr{P}}(p_1,...,p_{i-1},p_{i0},p_{i+1},...,p_n)$$
$$dp_1...dp_{i-1}dp_{i+1}...dp_n$$

These factors are essentially the statistical averages with all the parameter values changing according to their statistical descriptions, except the ith parameter, which is held constant. A plot of the statistical factor versus the independent parameter will graphically show the statistical dependence on the variable. If the factor is essentially flat as p_{i0} varies through the tolerance range, the average performance or yield is not sensitive to the ith parameter.

To improve calculating the average performance and yield factors, define the unbiased estimators $\hat{G}(p_{i0})$ and $\hat{Y}(p_{i0})$

$$\hat{G}(p_{i0}) = \frac{1}{M} \sum_{i=1}^{M} G(P_{i0})$$

$$\hat{Y}(p_{i0}) = \frac{1}{M} \sum_{i=1}^{M} \text{Accept}(P_{i0})$$

where P_{i0} is a parameter vector chosen according to $f_{\mathscr{P}}(P)\delta(p_i - p_{i0})$ (i.e., the parameters chosen according to the parameter density $f_{\mathscr{P}}(P)$ except the ith parameter which is held at p_{i0} and the density properly normalized).

We can further simplify the calculation of these factors by dividing the values of p_i into several equally sized regions called "bins." We then develop an approximation to the statistical factors by performing a Monte Carlo analysis in which all of the parameters are allowed to vary according to $f_\mathscr{P}(P)$ and the "binned" statistical factors for each parameter are evaluated separately for each bin. A plot of the binned approximation of the statistical factors versus the p_{i0} bin center values is called a *statistical factor histogram*. This technique is efficient to calculate because all statistical factors are determined with one M-point Monte Carlo simulation.

4.7.2 Average Performance and Yield Sensitivity Calculation

We can calculate a large-change statistical factor sensitivity as

$$S^{\bar{Y}}(p_{i0}) = \bar{Y}(p_{i0}) - \bar{Y}(p_{i0} + \Delta p_{i0})/\Delta p_{i0}$$
$$S^{\bar{G}}(p_{i0}) = \bar{G}(p_{i0}) - \bar{G}(p_{i0} + \Delta p_{i0})/\Delta p_{i0}$$

The interpretation of these sensitivities follows: $S^{\bar{Y}}(\Delta p_i)$ is yield calculated with all the parameters varying except the ith parameter fixed at p_{i0} minus the yield calculated with all the parameters varying except the ith parameter fixed at $p_{i0} + \Delta p_i$. This is not the same as calculating the yield with the parameter nominal value at p_{i0} and then at $p_{i0} + \Delta p_i$. The density changes from $f_\mathscr{P}(P)$ to $f_\mathscr{P}(P)\delta(p_i - p_{i0})$. The benefit to be gained from this is its ease of calculation, as shown next.

All the information for calculating the large-change statistical sensitivity can be obtained from the factor histograms. Let Δp_i be the extent of one or more of the bins in the factor histograms. Then $Y(p_{i0})$ is the value of the yield factor histogram at the bin covering the value p_{i0}, and $Y(p_{i0} + \Delta p_{i0})$ is the value of the yield factor histogram at the bin covering the value $p_{i0} + \Delta p_{i0}$. The yield factor statistical sensitivity is the slope of the yield factor histogram as illustrated in Figure 4.7.

Since these sensitivities are expressed in terms of the difference of two estimated values, the errors in this calculation can be large. A careful error analysis

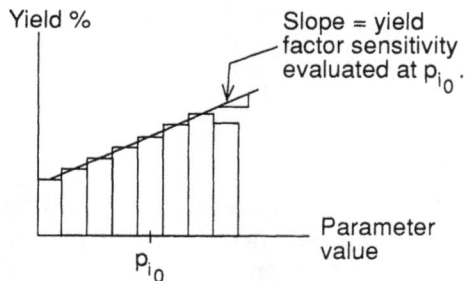

Figure 4.7 Yield factor sensitivity as the slope of the yield factor histogram.

is recommended. Typically 1000 points are used in the Monte Carlo simulation to determine the factor histograms. For nine bins, this gives on the average, over 100 points per bin. Assuming a 70% bin yield and a 95% confidence level, we find the yield error per bin is approximately ±9%.

With these kinds of errors, the numerical accuracy of the slopes need careful attention. We have fit lines through all the bins, using a least squares fit, and then assigned the line slope as the parameter sensitivity [11]. This has produced good results, although a complete error analysis is not presently available.

4.8 STATISTICAL SENSITIVITY MANAGEMENT AND REDUCTION

Sensitivity reduction is really the same as gradient optimization, because the sensitivity, as we have defined it, is simply the gradient of the performance (or performances) that is being targeted by the sensitivity figure. Standard single-point optimizers perform single-point sensitivity reduction; therefore, there is little or nothing to be gained by implementing a single-point sensitivity reduction study on a unit that has been single-point optimized.

The single-point sensitivities will not detect or correct the brinksmanship type of performance described in Chapter 1. As we have witnessed, brinksmanship performance can be bad for the unit's manufacturing yield. The sensitivity study that can be of use in determining the manufacturability of a unit must be at least a large-change sensitivity, and preferably a statistical-sensitivity study using the statistical-sensitivity figures introduced in this chapter. It is only through a statistical-sensitivity study that the designer can gain insight into how the actual parameter combinatorial behavior will affect the statistical performance of the design.

4.8.1 Sensitivity Management

In many instances, there are unit parameters which are well-controlled during the manufacture and lifetime of the unit. For example, the line widths on stripline microwave circuits can be manufactured and maintained with accurate dimensions. However, there are usually parameters which cannot be controlled or maintained, such as active device parameters, which for GaAs can vary ±20% or more. It is usually advantageous to determine a design such that the performance is sensitive to the parameters that are well-controlled, and not sensitive to the parameters that are not well-controlled. This we term *sensitivity management* because the sensitivity is being directed to a set of controllable parameters. This will, of course, reduce the performance variance when the parameter statistical variations are taken into account.

An idealized example of the use of sensitivity management is shown in Figure 4.8. The indicated acceptability region is always narrow in one dimension, thus

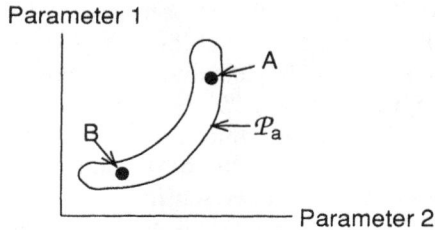

Figure 4.8 Idealized parameter space acceptability region, $\mathcal{P}_a$, which allows for sensitivity management. Design nominal parameter vector A, shows performance sensitivity to parameter 2, while design parameter vector B shows performance sensitivity to parameter 1.

indicating a statistical sensitivity when the parameter joint density is symmetric. If point A is chosen as the parameter nominal value, the design would be sensitive to parameter 2 because the extent of the acceptability region is narrow in the parameter 2 dimension. Conversely, if point B is chosen as the nominal parameter value, then the sensitivity is transferred to parameter 1. A good yield optimizer should be able to find and choose among the points A and B to maximize the yield, given the statistical distributions for the components.

The importance of the parameter distributions cannot be ignored in the sensitivity analysis, as illustrated in Figure 4.9. In this figure, the design point P_0 shows a sensitive design in both parameter 1 and parameter 2. If the parameters are uniform and independent, the parameter extents cannot practically exceed the extents labeled uncorrelated. If the parameters are correlated, with a correlation coefficient near 1.0, then the design is much less sensitive to the parameter values,

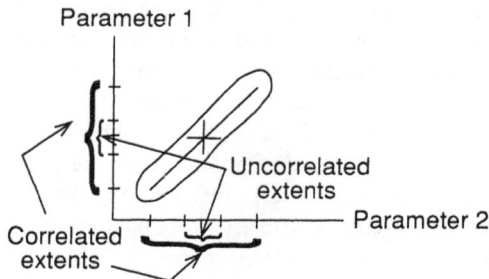

Figure 4.9 Idealized parameter space acceptability region, $\mathcal{P}_a$, which shows the effects of parameter correlations on the statistical sensitivity. For uncorrelated parameters, the sensitivity to the parameter variations is more than for parameters with a positive correlation coefficient.

and the parameter extents can be practically given by the extents marked corre-lated. This type of performance is common in filters, where performance is set by the ratio of two parameters.

4.8.2 Sensitivity Reduction

An example of sensitivity reduction is the performance variability reduction de-scribed in [12]. Performance variability reduction is similar to performance sen-sitivity reduction. The problem is posed in [12] as minimizing the variance, σ^2, of unit performance, G, over the tolerance region, T, subject to constraints on the yield and the tolerances.

The performance variance is estimated by a Monte Carlo simulation

$$\hat{\sigma}^2 = \frac{1}{M-1} \sum_{i=1}^{M} \left\{ G(P_i) - \frac{1}{M} \sum_{i=1}^{M} G(P_i) \right\}^2$$

and the yield is again estimated by the same Monte Carlo simulation given by

$$\text{Yield Estimate} = \hat{Y} = \frac{1}{M} \sum_{i=1}^{M} \text{Accept}(P_i)$$

where in both cases the P_i are chosen according to the parameter joint density $f_{\mathscr{P}}(P)$.

A given iteration of the variability-reduction algorithm proposes that

1. The responses G(P) and the response single-point sensitivities, S_{pi}^G are cal-culated for each of the M points in the Monte Carlo simulation.
2. A search direction S^r, with r denoting the rth iteration, is chosen that reduces the performance variance. This direction is modified if it indicates that the yield will be reduced below a predetermined lower limit.
3. A step length is calculated so as to minimize the variance, and a new nominal parameter value is chosen.

In the cited algorithm, a quadratic approximation to the performance variance is made. This simplifies the determination of the step direction. The step size is determined by either making a linear approximation to G(P) around P_0, or doing a search along the step direction until the variance estimate is minimum or the yield constraint is violated.

The results of this variance-reduction technique are demonstrated on a seven-parameter second-order active RC filter with good results. This type of algorithm is useful because it makes full use of the statistical parameter variations in deter-mining sensitivity and its minimization.

4.9 EXAMPLES

To better illustrate the statistical sensitivities, statistical factors, and estimates that were presented in this chapter, two examples are presented: the lug nut and a Salen and Key filter. We have determined the yield and performance factors for each case and each will be discussed.

4.9.1 Lug Nut

The lug-nut problem considered here has four parameters (p_1, p_2, p_3, p_4). The performance, W, is written in terms of the parameters as

$$W = \text{minimum}(p_1, p_2, p_3, p_4)$$

There is symmetry in this problem in that all the parameters appear in the mathematical expression for W in the same way. This is useful because when the nominal value is the same for all parameters, the performance factors and sensitivities will be the same for all parameters.

First, we determine the statistical model. For this example we make all the parameters uniformly distributed, independent, and varying over the interval $0 \leq p \leq 1$. In keeping with the lug-nut spirit of this problem we will assign the nominal value to be $(0,0,0,0)$.

Next, we performed a Monte Carlo yield calculation on this problem, with the performance specification $0 \leq W \leq 0.1$. The yield estimate, using a 10,000 trial Monte Carlo simulation, is $Y = 34.3\%$ with a 95% confidence interval of less than $\pm 1\%$. The theoretical yield is 34.30%. (For the interested reader, see Exercise 4.2.)

We then divided each parameter into ten regions or bins. For each bin we estimated several factors: the yield factor, the average performance factor, and the performance-variance factor. Each of these is plotted in Figures 4.10, 4.11, and 4.12, respectively. Each of these plots is made at the nominal value $(0,0,0,0)$.

A quick look at Figures 4.10, 4.11, and 4.12 shows the effectiveness of these statistical factors, because each indicates the need for the parameter to be set at zero for good statistical performance.

If the statistical sensitivities are needed for this problem, they can be estimated by taking the slope of the appropriate factor plot, using a large-change type of calculation. This problem indicates clearly that statistical (or classical) sensitivities do not in themselves provide total information. For instance, the yield sensitivity is practically zero for the parameters above 0.1, yet the yield is also lowest for these parameter values. Clearly there is a need to minimize sensitivities subject to some limits on desirable performance or yield.

Yield Factor

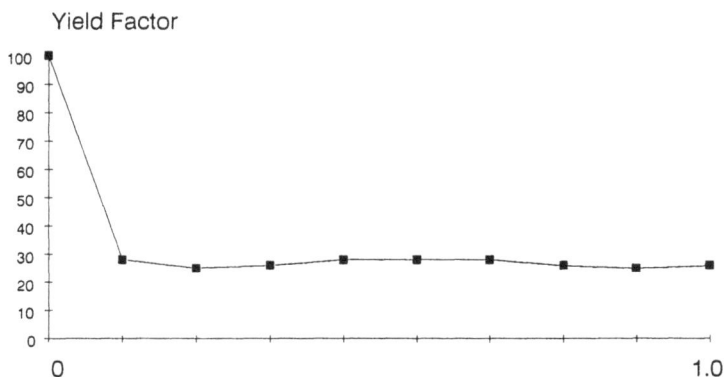

Figure 4.10 Plot of the estimated yield factor, in percent, versus parameter value for the lug-nut problem, evaluated at $P_0 = (0,0,0,0)$.

Average Performance Factor

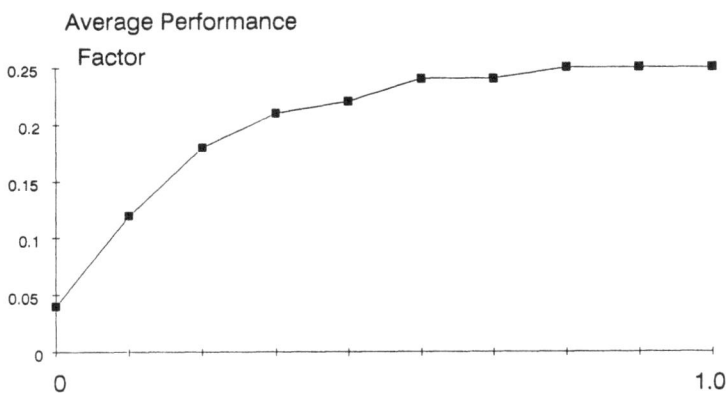

Figure 4.11 Plot of the estimated average performance factor versus parameter value for the lug-nut problem, evaluated at $P_0 = (0,0,0,0)$.

Another measure that has been used lately is a "signal-to-noise" measure such as the average performance divided by the performance standard deviation. The approach is to integrate good and bad characteristics into a single metric, which can be maximized (or minimized in this case) using optimization. For this problem we have plotted in Figure 4.13 the average performance mean factor divided by the standard deviation of the performance factor. This was obtained by dividing the appropriate factors displayed in the previous variance and average

Variance Factor

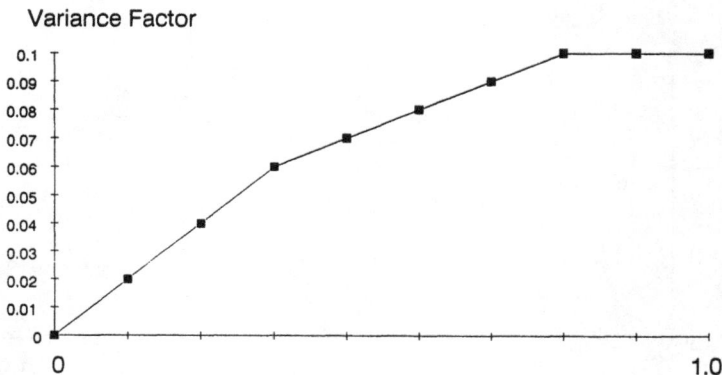

Figure 4.12 Plot of the estimated performance-variance factor versus parameter value for the lug-nut problem, evaluated at $P_0 = (0,0,0,0)$.

Average Performance
―――――――――――
Standard Deviation

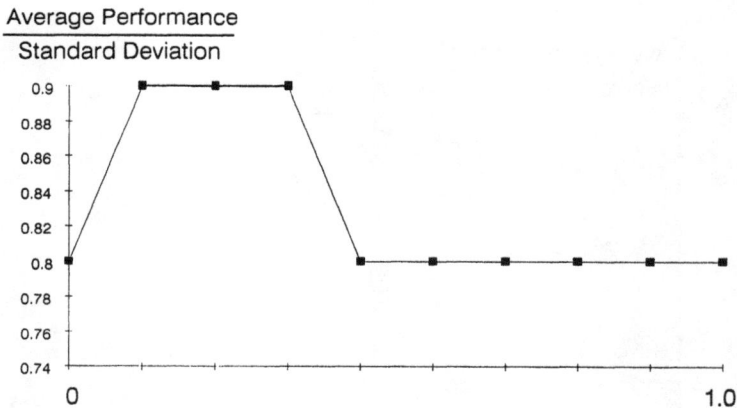

Figure 4.13 Plot of the estimated performance mean divided by the performance standard deviation factors versus parameter value for the lug-nut problem, evaluated at $P_0 = (0,0,0,0)$.

performance curves. As seen from this curve, it is essentially flat for this problem. This signal-to-noise property is used in most of the Taguchi Methods that have been presented [2].

4.9.2 Salen and Key Filter

The next example involves a Salen and Key lowpass filter which is used often in the statistical-design literature [13]. The circuit configuration is shown in Figure 4.14, and the nominal parameter values and their tolerances are given in Table 4.3.

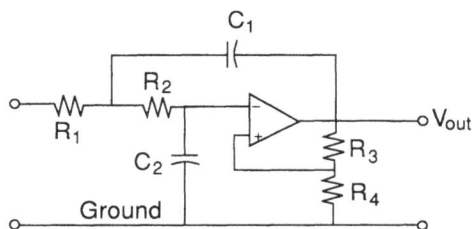

Figure 4.14 Salen and Key filter configuration.

Table 4.3
Nominal Values and Tolerances for the Salen and Key Filter

Parameter	P_0	Tolerance
R1	1 kΩ	±1%
R2	2.5 kΩ	±1%
R3	4 kΩ	±1%
R4	12 kΩ	±1%
C1	1 μF	±0.02%
C2	0.1 μF	±0.02%
A	3000	none

The filter Q is used as the performance, and it is mathematically described as

$$Q = \frac{(R_1 R_2 C_1 C_2)^{-1/2}}{\left(\dfrac{1}{R_1 C_1} + \dfrac{1}{R_2 C_1} - \dfrac{R_3}{R_2 R_4 C_2}\right)}$$

The Q at the nominal parameter values is 19.0. The statistical model uses uniform independent parameters with the indicated parameter tolerances. The specification for the circuit is $10 \le Q \le 40$. The estimated yield is 88.7% ±1% at a 95% level of confidence, by way of a 10,000 trial Monte Carlo analysis..

At the nominal value given in Table 4.3, we calculated the yield factor, the average performance factor, and the performance standard deviation factor. Each of these is given in Figures 4.15, 4.16, and 4.17 respectively.

These factor plots are more complicated and challenging to understand than the ones for the lug-nut problem because each parameter has a different effect on the factors.

The yield factor plot in Figure 4.15 indicates that if R_3 is held constant, and all the other parameters are allowed to vary, then the yield increases with a decrease in R_3. Notice that with R_3 fixed at the nominal value, the yield is about the same

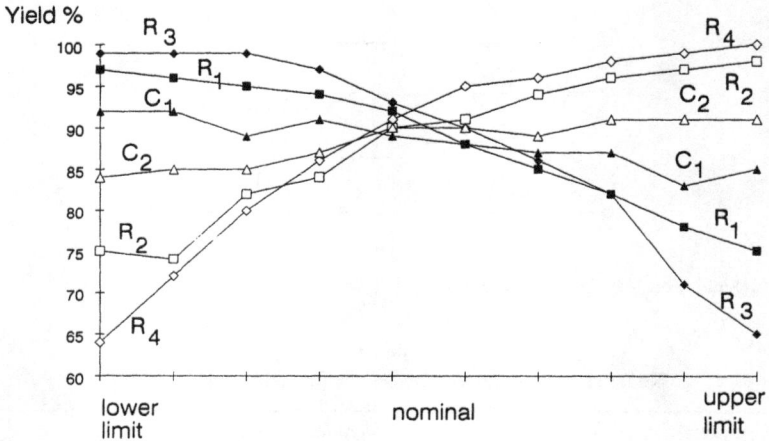

Figure 4.15 Plot of the estimated yield factor versus parameter value for the Salen and Key filter problem, evaluated at P_0 given in Table 4.3.

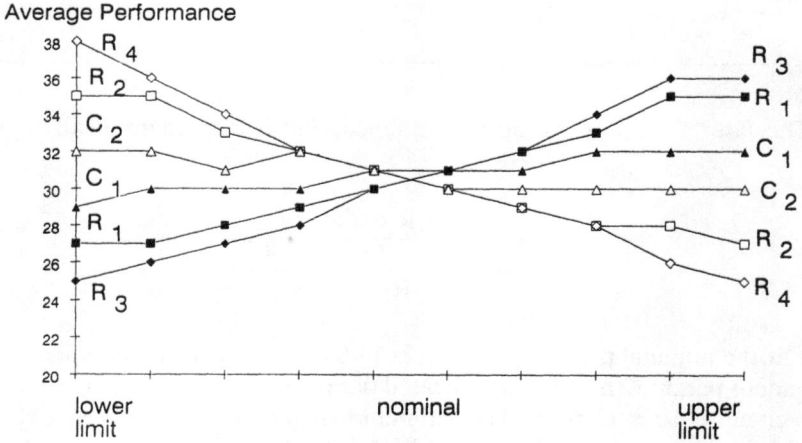

Figure 4.16 Plot of the estimated average performance factor versus parameter value for the Salen and Key filter problem, evaluated at P_0 given in Table 4.3.

as with R_3 varying. Similar statements about the statistical behavior of this filter can be made for the other parameters too. The yield is least sensitive to the value of C_1, while it is most sensitive to the values of R_3 and R_4. From this yield plot we might expect the yield to rise if we either reduce R_3 or increase R_4. Remember that the yield with the parameters at their nominal value is 89% ±1%. Having

Standard Deviation Factor

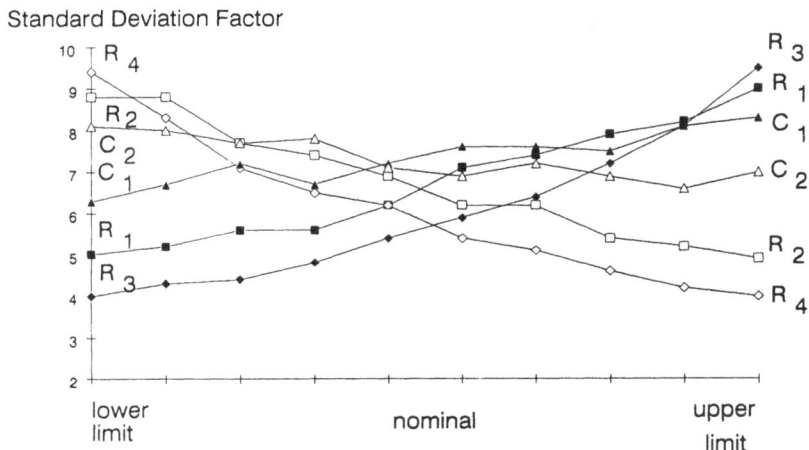

Figure 4.17 Plot of the estimated performance standard deviation factor versus parameter value for the Salen and Key filter problem, evaluated at P_0 given in Table 4.3.

performed this experiment using 10,000 Monte Carlo simulations (95% confidence less than $\pm 1\%$), we find that the yield is 96% with R4 raised by 0.5%, and 96% with R_3 lowered by 0.5%. Also, the yield does not significantly change when C_1 is increased or lowered by 0.1% (87% with C_1 increased and 90% when C_1 is decreased). The essence of statistical design is the fact that by changing either R_3 or R_4 the yield is increased.

The average performance factor shows the inverse of the yield factor trends. We also see a disturbing trend. Although the nominal value (single-point) Q is 30.0, the average Q is around 31.2. This indicates that the performance variance is large and that the statistical variations tend to raise the Q on the average. This is seen in the standard-deviation factor, where the standard deviation of the performance varies between 4 and 10, depending on the values of the components. We also see that increasing R_4 or decreasing R_3 decreases the performance standard deviation, a very good result, as performance variance is usually best when minimized.

The last factor examined is signal-to-noise factor composed of the ratio of the average performance factor and the standard deviation factor plotted in Figure 4.18. The signal to noise is improved along with yield by increasing R_4 or decreasing R_3.

4.10 CONCLUSION

Sensitivity studies are crucial to the design of reliable and manufacturable units. Both the designer and the manufacturing engineer must know which parameters

Average Performance
Standard Deviation

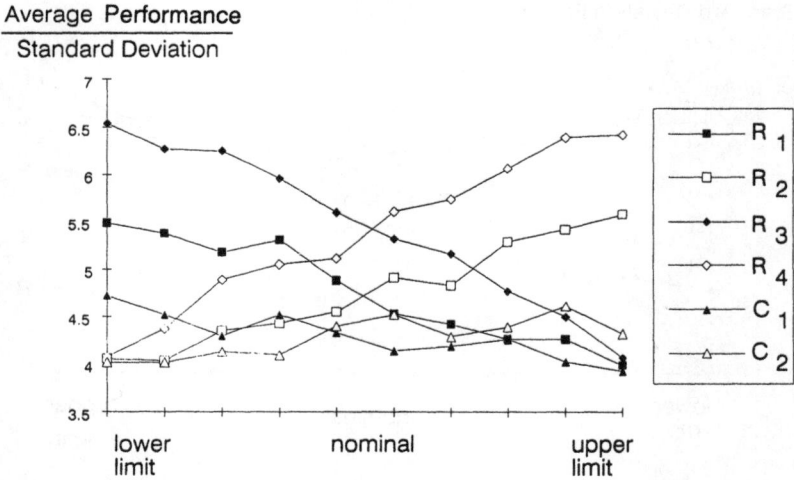

Figure 4.18 Mean divided by standard-deviation factor for the Salen and Key filter evaluated at the nominal parameters given in Table 4.3.

strongly affect performance. These parameters must be closely controlled during manufacture or possibly tuned during testing. Recently published works have concluded that a "quality" design exhibits "small" performance variation during manufacture and lifetime operation. The key to understanding performance variance and designing for performance-variance minimization lies in the calculation and understanding of unit sensitivities.

Classical sensitivities, as presented in the first part of this chapter, do not fully represent the variation mechanisms that the unit will encounter during manufacture or lifetime operation. Classical sensitivities are developed using small changes in the parameters, and changing only one parameter at a time. In reality, units will undergo large changes in parameters, and the parametric changes will be simultaneous, governed by the parameter joint probability density function (JPDF). To overcome these deficiencies in the classical sensitivities, statistical sensitivities are introduced. Much of the work presented here on statistical sensitivities is new.

To understand the need for statistical sensitivities, a clear understanding of the unit parameter, P, is needed. The unit parameter is the basic element, the design language if you wish, of unit design as it is now undertaken. Classical sensitivity and design assumes that the design parameter is P, and that the performance is given by G(P). The classical-design problem then becomes determining the best value for P, based on the performance G(P). Using P as the design parameter implies the designer can set and control the exact parameter values that will be encountered during manufacture and unit lifetime. This is not the case. In reality P can, and must, be broken down into (at least) two components; $\mathcal{P} = P_0 + \Delta\mathcal{P}$, where P_0 is the parameter average or nominal value, and $\Delta\mathcal{P}$ is the statistical variation on P. Statistical sensitivity and design take the viewpoint that P_0, the parameter average value, is the design parameter available to the designer, and $\Delta\mathcal{P}$ is set by the unit's manufacturing and lifetime environment. When employing the statistical approach (i.e., using P_0 as the design parameter), the designer chooses the best P_0, which results in a family of unit parameters, $\mathcal{P} = P_0 + \Delta\mathcal{P}$, which then gives rise to a family of performances $G(P_0 + \Delta\mathcal{P})$. Managing and analyzing this family of parameters and performances, as a function of P_0, is the significant accomplishment of the statistical sensitivities.

We have presented a way of estimating the statistical sensitivities based on the average performance and yield factors. These can be calculated for all unit parameters using one M-point Monte Carlo simulation. The lug-nut and Salen and Key filter examples illustrate the usefulness of the statistical sensitivities, as well as illustrate the advantage they give the designer over the classic sensitivities.

4.11 IMPORTANT IDEAS FROM CHAPTER 4

Section 4.1

- Sensitivity studies are important to the designer, as they give the relation between parameters and performance, indicating how parameter variation affects performance variation.
- Sensitivities are important in optimization techniques.
- Reducing the sensitivity of a design can partially increase a design's quality.
- We introduce three sensitivity definitions:
  1. Manufacturing sensitivity is an actual change in manufacturing statistical performance as a function of a change in a design parameter value.
  2. Classic sensitivity is the change in calculated performance as a function of a change in a parameter value.
  3. Statistical sensitivity is the change in a calculated performance statistic as a function of a change in a design parameter value.

Section 4.3

- There are two important properties that should be included in a sensitivity study:
  1. Large changes in the parameters;
  2. Parameter variations as defined by the parameter joint PDF.
- Classic sensitivities alone do not necessarily include both of the above desirable properties.

- Taguchi Methods can do a good job of interrogating the performance space, and they can be very economical in using experimental data. But remember, you get what you pay for, and Taguchi methods do not accommodate the parameter joint PDF. Further they require a lot of user intervention and "steering."

Section 4.4

- Statistical sensitivities, as introduced, are recommended for determining the sensitivity to manufacturing and lifetime parameter variations.

Sections 4.5 and 4.6

- It is useful to look at performance variance as a function of the design parameters P_0. One way to accomplish this is with a statistical factor histogram.

Section 4.7

- Performance and yield factors are useful for calculating the statistical sensitivities. These can be calculated for all parameters by performing one M-point Monte Carlo simulation.

Section 4.8

- Sensitivity management is an important requirement of all unit designs because it addresses the question of unit reliability and quality.

Section 4.9

- The Salen and Key filter and lug-nut problems illustrate the effectiveness of the statistical sensitivities developed in this chapter. The lug-nut problem, although very nonlinear, shows the shortcomings of the classic sensitivities.

EXERCISES

Exercise 4.1

You might want to try evaluating $\bar{W}$ for different relationships on p_{10} and p_{20}. For instance, you might verify that when $p_{20} \geq p_{10} + 0.1$, $\bar{W}(p_{10}, p_{20}) = p_{10} + 0.1^2/2$.

Exercise 4.2

Write a Monte Carlo yield-calculation program for the lug-nut example problem. The performance specification is $0 \leq W \leq 0.1$; the yield estimate, using a 10,000 trial Monte Carlo simulation, is $Y = 34.3\%$ with a 95% confidence interval of less than $\pm 1\%$; the theoretical yield is 34.30%. Because the mathematics relating W and the parameters are so simple here, this is an ideal problem to start programming. Let the number of trials in the Monte Carlo simulation be an input variable, and see how the yield estimate changes as the number of trials change. It shouldn't take more than 20 lines of code to do this calculation.

REFERENCES

[1] R. Aguayo, *Dr. Deming, The American Who Taught the Japanese About Quality*, Carol Publishing Group, A Lyle Stewart Book, 1990.

[2] M.S. Phadke, *Quality Engineering Using Robust Design*, Englewood Cliffs, NJ: Prentice-Hall, 1989.

[3] J. Vlach and K. Singhal, *Computer Methods for Circuit Analysis and Design*, New York, NY: Van Nostrand Reinhold, 1983.

[4] J. Purviance and M. Meehan, "A Sensitivity Figure for Yield Improvement," *IEEE Trans. on Microwave Theory and Techniques*, Vol.36, No. 2, Feb. 1988, pp. 413-417.

[5] G. Taguchi, *Introduction to Quality Engineering, Designing Quality into Products and Processes*, Tokyo, Japan: Asian Productivity Organization, 1986.

[6] M.S. Phadke, R.N. Kackar, D.V. Speeney and M.J. Grieco, "Off-line Quality control in Integrated circuit Fabrication Using Experimental Design," *AT&T Technical Journal*, 1983.

[7] K.M. Lin and R.N. Kacker, "Optimizing the Wave Soldering Process," *Electronic Packing and Production*, Feburary 1986.

[8] R.N. Kackar and A.C. Showmaker, "Robust Design: A Cost Effective Method for Improving Manufacturing Processes," *AT&T Technical Journal*, 1986.

[9] W. Nazaret and W. Klingler, "Tuning Computer Systems for Maximum Performance: A Statistical Approach," *The American Statistical Association*, 1986.

[10] K. Dehnad, *Quality Control, Robust Design, and the Taguchi Method*, Pacific Grove, CA: Wadsworth and Brooks/Cole Statistics/Probability Series, 1989.

[11] J. Sarker and John Purviance, "Yield Sensitivity of AlGaAs High Electron Mobility Transistor," *Int. J. of Microwave and Millimeter-Wave Computer-Aided Engineering*, Vol. 2, No. 1, pp. 12-27, Jan. 1992.

[12] A. Ilumoka, N.Maratos, and R. Spence, "Statistically Based Algorithms for the Reduction of Circuit Performance Variability," *Proc. of the Int. Symp. on Circuits. and Systems*, CAS - 1981, pp 149-152.

[13] R. Spence and R. Soin, *Tolerance Design of Electronic Circuits*, Reading MA: Addison-Wesley, 1988.

Chapter 5
Yield Optimization

"For the circuit designer, electrical performance is no longer of overriding interest. Rather, the goal is the best design in terms of total quality. By total quality is meant a balance of performance, ease of manufacture, and ease of support." [1]

"There is a need to optimize the unit performance, not at a single parameter value, but over the entire range of parameter values that will be encountered during manufacture."

5.1 INTRODUCTION

So far this book has presented yield sensitivity and the calculation of yield. In this chapter, yield optimization, which is the focus of statistical design, is discussed. Yield optimization rather than performance optimization is an important goal for a manufacturable design. But since yield is often impossible to calculate exactly, optimizing yield is not a straightforward task.

First, an overview of optimization including definitions and results required in subsequent sections is presented. Because yield optimization can be viewed as a natural extension to the nominal optimization process, we present an overview of the single-point optimization strategy. Finally, yield optimization ideas are discussed along with reviews of several practical statistical-design techniques of today.

5.1.1 The Optimization Problem

With the advent of high-speed engineering-workstation computers and a commensurate increase in the complexity of the design problem, optimization has become a necessary part of complex unit design, as in microwave circuits and systems. The overall design process using optimization is illustrated in Figure 5.1, where the unit is first conceived by the designer, then analyzed and improved (i.e. optimized) until the performance is either acceptable or the best that can be realized. Generally computer algorithms are responsible for the decision-making inside the "modify unit parameters" block of the optimizer. The unit performance can be improved by changing many different properties of the unit, like its structure or the technology used to implement the unit, or the unit physical configuration, but the state of the art in optimization involves adjusting the unit parameters without changing the structure, technology, or physical considerations. This procedure is referred to as *parametric optimization* and is the simplest and most mathematically complete form of the optimization problem. Fully automated optimization of unit structure or technology is yet to be accomplished by the design community.

The formulation of the parametric optimization process is as follows: Given a single-valued objective function,

$$O(p_1, p_2, \ldots, p_n) \tag{5.1}$$

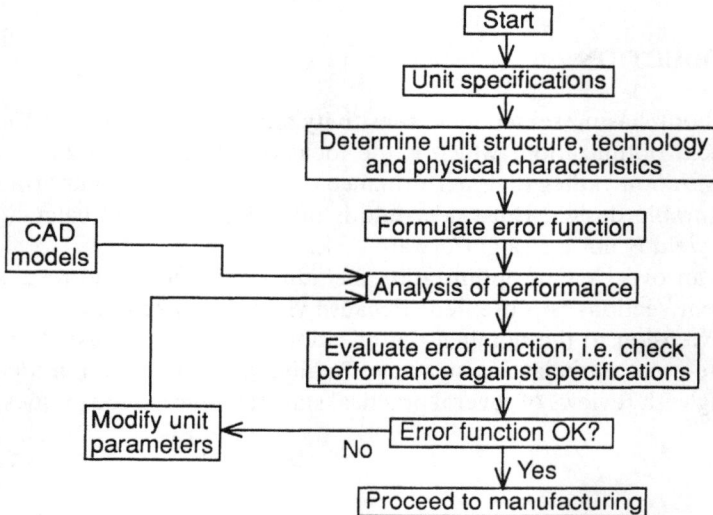

Figure 5.1 The design and optimization of complex units using computer-aided optimization.

the unconstrained optimization problem is stated as minimizing the objective function

$$\min_{P} O(P)$$

where P is a member of parameter space, $\mathcal{P}$.

There are often constraints placed on the design parameters as well, usually corresponding to some physically meaningful restriction, (i.e., positive values for length and resistance)

$$p_1 \geq 0, \ p_2 \geq 0,...,p_n \geq 0.$$

Minimization of O(P) subject to the constraint equations is called *constrained optimization*.

The general approach taken by modern optimization techniques is to create a sequence of parameter vectors $P_1, P_2,...P_k$, such that [2]

$$O(P_1) > O(P_2) > ... > O(P_k) \tag{5.2}$$

This sequence, if found, will result in the *absolute minimum* of the objective function, provided it is convex (see Section 3.5.2). The parameter vector which minimizes O(P) is then called the *global minimum, P**. If O(P) is not convex then this sequence may terminate at a *relative (or local) minimum*. Figure 5.2 shows some one-dimensional objective functions. Curve A is convex and thus the sequence given in (5.2) will terminate (i.e., converge) to the global minimum, re-

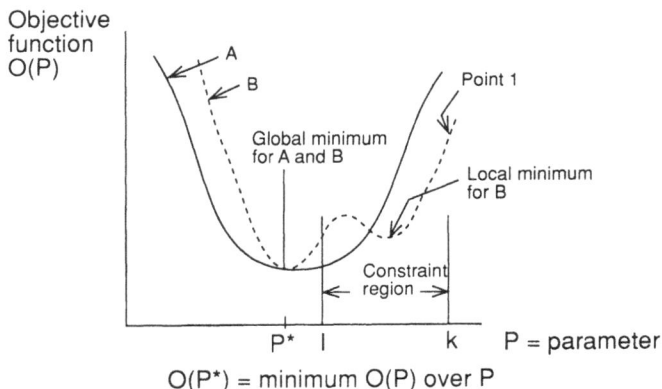

$$O(P^*) = \text{minimum } O(P) \text{ over } P$$

Figure 5.2 Examples of (a) convex and (b) nonconvex objective functions, and a constraint region for a one-dimensional objective, O(P).

gardless of the initial starting point. Curve B illustrates a nonconvex objective function; if the starting point of the sequence is point "1", the sequence may end up in the relative minimum rather than in the global minimum.

After a minimum is found by the optimizer, it remains to be classified as global or relative. This distinction can be difficult to make. The problem of an optimizer terminating at a relative minimum is sometimes referred to as *false termination*. (There are optimization algorithms, specifically "random" optimizers which address false termination, and more will be said about them later in this chapter.)

A practical way to determine if false termination has occurred is to initiate optimization using different initial parameter vectors. If each converges to the same point, it is likely to be the global minimum. But if each converges to a different minimum, the objective function is probably not convex and the solution vector having the smallest value for the objective function should be used.

As an example of constrained optimization, assume that the constraint equation is $l < P < k$. With this constraint, which is included in Figure 5.2, the correct solution is the relative minimum because the global minimum lies outside the constraint range.

5.1.2 Classification of Optimization Methods and Goals

There are numerous ways to formulate the objective function (5.1) and to choose the sequence given in (5.2). For performance optimization, the objective is a function of unit performance; for yield optimization, the objective is a function of unit yield (or yield estimate). The sequence (5.2) can be determined by gradient methods or by direct-search techniques. These apply for both yield and single-point performance optimization. A classification of the optimization options is given in Table 5.1.

Table 5.1
Minimization Techniques and Objective Function Formulations for Single-Point and Statistical Optimization

Objective function	Minimization techniques
single-point optimization	quasi-Newton
sum of squares	conjugate gradient
minimax	Gauss-Newton
sum of pth power	combined methods
sum of absolute value	simplex/complex
least p^{th} and generalized least p^{th}	stochastic gradient
statistical optimization	genetic
yield	center of gravity (statistical only)
average performance	
margin sensitivity	
performance variance	
multicircuit	

Minimization techniques and objective function formulations can generally be mixed. For instance, it is possible to use a random technique to minimize the sum of the absolute errors, or to use a random technique to minimize statistically the average performance, or to maximize the yield. Some of the combinations such as gradient minimization of the single-point performance error are popular, while other combinations are not fully understood and tested, like the genetic minimization of the average performance error. There is much work to be done in classifying and understanding the benefits to each of these optimization possibilities.

Since single-point optimization is usually best understood and has strong ties to yield optimization, we next examine the single-point optimization strategy.

5.2 SINGLE-POINT (NOMINAL) PERFORMANCE OPTIMIZATION

Single-point optimization is the determination of a single parameter vector, P^*, such that $O(P^*)$ is an absolute minimum:

$$O(P^*) < O(P)$$

for $P \in \mathcal{P}$ and $P^* \in \mathcal{P}$. This is called nominal optimization, because the optimum parameter vector, P^*, is chosen as the nominal value for the parameters when the unit is manufactured. The initial step in the optimization process is to decide which objective function formulation to use. This formulation can have a strong effect on the outcome of the optimization, and therefore it deserves to be discussed here and carefully considered in the design process.

5.2.1 Objective (Error) Function Formulation

The objective function for single-point optimization is often formulated as a summation of fundamental terms called *residuals*. A residual is simply the weighted difference between the unit performance for measurement i, and the corresponding specification. Mathematically, this is expressed

$$
\begin{aligned}
r_i &= w_i(m_i - s_i) \\
r_i &= w_i|m_i - s_i| \\
r_i &= w_i(s_i - m_i)
\end{aligned}
\tag{5.3}
$$

for an upper bound specification, for an equality specification, and for a lower bound specification, respectively, where w_i is a positive weighting factor, m_i, is the ith unit performance measurement (i.e., $m_i = g_i(P)$), and s_i is the specification for the ith measurement. The role played by the w_i is twofold: first to compensate for differences in units between measurements, and second to allow emphasis to be placed on certain "important" residuals.

Notably, (5.3) supports the convention that $r_i > 0$ is indicative of "violation of specification" or "error." For this reason the combination of residuals is commonly referred to as the *error function*. For example, the classical least squares error function is given as the sum of residuals squared

$$EF = \sum_{i=1}^{k} r_i^2, \text{ for } r_i > 0 \qquad (5.4)$$

Since only positive-valued residuals enter into the formulation, there is a lower bound on the least squares error function of zero; optimization stops when all specifications have been satisfied.

While terminating optimization with all residuals less than or equal to zero sounds ideal, Bandler and Charalambous [3] have gone an important step further and proposed the *generalized least p^{th}* error function formulation:

$$EF = \left(\sum_{r_i > 0} r_i^p \right)^{1/p} \qquad \text{if } r_{max} > 0$$

$$EF = 0 \qquad \text{if } r_{max} = 0 \qquad (5.5)$$

$$EF = \frac{-1}{\left(\sum_{i=1}^{k} \left(\frac{-1}{r_i} \right)^p \right)^{1/p}} \text{ if } r_{max} < 0$$

where r_{max} is the largest (maximum) residual. Unlike conventional least pth formulations, (5.5) allows optimization to continue even after all specifications have been satisfied. Formulation (5.5) takes the designer's goals and attempts to "oversatisfy" them as much as possible, and as such, the optimization process requires less designer intervention and "coaxing."

The exponent p in (5.5) can be any positive number but usually appears as an even integer. The effect that p has on the error-function contour is best described by example. Suppose we have three residuals with values $r_1 = 0.5$, $r_2 = 1$, and $r_3 = 1.5$. Then the following Table 5.2 can be composed, where the residuals are listed as functions of p. The observation here is, of course, that the larger p, the more emphasis is placed on the largest residual. And as p tends toward infinity, only the largest residual contributes to the overall error function. This is the *minimax* error function.

The minimax error function tends to promote an "equal-ripple" or "Chebyshev" solution to a given set of unit specifications, where all residual magnitudes tend to be the same. In practice, the minimax solution can be obtained by *sequential least p^{th} optimization*, where sequential application of (5.5) with increasing values for p (i.e., $p=2 \rightarrow p=4 \rightarrow p=8 \rightarrow p=16$) is used.

The choice of the error function will determine the "error surface" in performance space that is to be minimized. Simple error functions may create error

Table 5.2
Residual Values as a Function of the Exponent p of the Least-pth Error Function

p	$(r_1)^p$	$(r_2)^p$	$(r_3)^p$
1	0.5	1	1.5
2	0.25	1	2.25
4	0.063	1	5.063
6	0.0156	1	11.39
20	0.9e-9	1	3325

surfaces that can be easily minimized. Likewise, complicated error functions can create error surfaces so complex as to seriously challenge modern optimization algorithms. The choice of error function is a critical part of single-point optimization. Care must be taken to subsume the designer's intent for optimal performance, but have an error function sufficiently simple that minimization algorithms will work. Much work needs to be disseminated to the general design community regarding the proper development of optimization error functions. Perhaps the relatively new field of visualization techniques could be employed to aid in this effort. As it stands, the choice of error-function formulation is largely empirical.

Assuming that a suitable error-function formulation has been determined, there are two main classes of search strategies that can be used to minimize the error: gradient and direct-search methods.

5.2.2 Gradient Methods

There are four broad categories of gradient-based minimization methods: Gauss-Newton, quasi-Newton, conjugate gradient, and composite methods (combinations of the other three). Algorithms within each category are delineated by the mathematical steps taking place for a given iteration from P_k to P_{k+1}. Generally, each iteration consists of solving a linear or quadratic programming problem to identify the next parameter vector of the error-function-reducing sequence $\{P_k\}$.

Because there is an abundant supply of literature on this subject, our goal here will be to summarize the practical aspects of some of the popular gradient-based minimization methods. (Some of the many references available are [2 and 4–9].)

Gauss-Newton Methods

Gauss-Newton methods are characterized by a linear approximation to the functional dependence each residual has to the unit input design parameters. In addition, so-called *trust regions* are used to limit the extent of parameter space over which the linear model is believed to be accurate. Heuristic schemes are typically used to dynamically regulate the size and shape of the trust region in an effort to improve the rate of convergence. While Gauss-Newton methods have been observed to provide fast initial convergence, their ability to finish can be quite abysmal. For this reason, researchers have augmented the first-order Gauss-Newton methods with second-order quasi-Newton methods, switching from one to the other based on dynamic rate of convergence information.

Quasi-Newton Methods

The hallmark of the quasi-Newton minimization category is the use of approximate second-order information combined with first-order derivatives and line searches. The second-order information provides reliability and rapid convergence in the neighborhood of a minimum. Moreover empirical evidence suggests that these methods are very forgiving with respect to "exactness" of the line search, especially when compared to the conjugate gradient methods.

Conjugate Gradient Methods

The main advantage of conjugate gradient minimization methods lies in their simplicity and minimal storage requirements on the computer. For huge optimization problems involving thousands of variables, the conjugate gradient methods are usually first to be deployed.

A requirement for all gradient-based optimization procedures are the first-order derivatives of the error function with respect to the variables. The gradient of the error function with n variables is defined as

$$\nabla EF(P) = (\partial EF/\partial p_1, \partial EF/p_2,...,\partial EF/\partial p_n)$$

Minimization continues until the gradient is zero (i.e., $\nabla EF(P^*) = (0,0,...,0)$) signifying a minimum (possibly relative) of the EF has been found. This solution is usually referred to as a *stationary* point.

Because analytic derivatives may not be available, the error-function gradient is usually approximated using finite differences:

$$\partial EF/\partial p_i \approx [EF(P + \Delta p_i) - EF(P)]/|\Delta p_i| \tag{5.6}$$

where $\Delta p_i = (0,0,...,\Delta p_i,...,0)$. A more accurate calculation is given by

$$\partial EF/\partial p_i \approx [EF(P + \Delta p_i) - EF(P - \Delta p_i)]/|2\Delta p_i| \tag{5.7}$$

but at the cost of an additional unit simulation $G(P_k)$. Note that using gradient approximation (5.6) requires $(n + 1)$ unit simulations $G(P_k)$. For problems with a large number of parameters, this expense can be prohibitive. Therefore, a number of methods have been proposed for function minimization that do not require gradient information: the *direct-search* methods.

5.2.3 Direct-Search Methods

Because conventional gradient-based minimization methods are driven by first-order derivative information, they are adept at quickly finding the nearest minimum (or maximum). Most of these techniques are locally driven and may not attempt to discover the global solution. One group of methods that attempts to search the performance space for the global minimum are called statistical or random-search methods.

The crudest form of random-search method would be to randomly choose a set of parameters, and then evaluate the performance. If the performance is the best so far, retain the parameters as the new best solution. This random selection process continues until a satisfactory performance is obtained or until the entire space has been searched. A flow diagram illustrating a crude random search is shown in Figure 5.3. This seemingly naive approach to optimization is very robust. It guarantees the discovery of a solution near the global optimum for any performance surface. And in the limit as the number of search points goes to infinity, the global optimum will be found.

Of course, this simple search algorithm is often impractical because of the computation time involved. But the robustness of the concept has intrigued researchers for years, where their focus has been to develop some kind of adaptive random search that preserves robustness yet uses information gathered from the

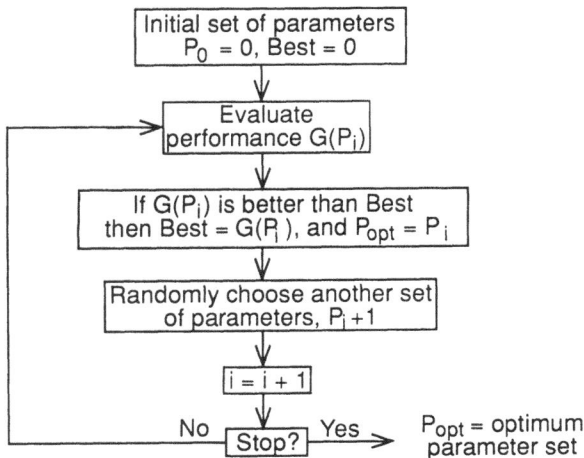

Figure 5.3 Flow diagram of a crude random-search optimizer.

past search to more carefully choose the points used in the future search, toward the goal of boosting efficiency. For instance, if every time negative parameters are used, the performance is poor, a "smart" search algorithm will eventually reduce the probability that negative parameters are chosen for the search.

Two forms of "smart" random searches are the random search using *simulated annealing* [10] and the *genetic algorithm* [13, 14]. These techniques are very powerful and although they have not been fully examined for the yield-optimization problem, their application is certain to be exploited in the near future.

Random Search Using Simulated Annealing

Simulated annealing takes its name from the thermodynamic properties of materials as they cool. As a crystalline material cools, it starts in a hot and energetic state. The atoms are excited and in violent motion. As the material cools, the atomic motion decreases and the material assumes a lower energy state. Finally, as the material solidifies, the atoms assume an energy state that has been minimized through this cooling process. For crystalline materials, the crystal is the minimum energy state, which is always "found" by the cooling system. The key to this system of decreasing energy is that temperature is slowly reduced so that the material is always in a steady state of thermal equilibrium.

The application of these ideas to random optimization is intriguing. The analogy of the "temperature" of the algorithm represents the extent of the random search. Metropolis et al. [11] introduced a method where, at each step of the algorithm, the present "best" parameter set is randomly perturbed, with the perturbation proportional to the "temperature" of the algorithm. The change on performance, ΔP, is determined. If the ΔP is negative (i.e., the perturbed performance is less than the present "best" performance) the perturbed point is rejected. If the ΔP is positive then the Boltzman probability

$$\text{Prob}(\Delta P) = e^{-\Delta P/kT} \tag{5.8}$$

is calculated, where k is Boltzman's constant, and T is the algorithm "temperature." A uniform random number chosen from the interval [0,1], is then generated and compared to $\text{Prob}(\Delta P)$. If $\text{Prob}(\Delta P)$ is greater than the random number, then the new perturbed point is retained as the "best" point, otherwise the perturbed point is rejected. By repeating these basic steps, one simulates the thermal motion of atoms at temperature T. The system temperature, T, is gradually decreased until it becomes zero, with the method terminating with the present "best" parameters at the resulting optimum. Figure 5.4 illustrates this algorithm for simulated annealing.

This search procedure is generally as robust as the random search procedure, assuming the temperature is not dropped too quickly, but decreasing temperature

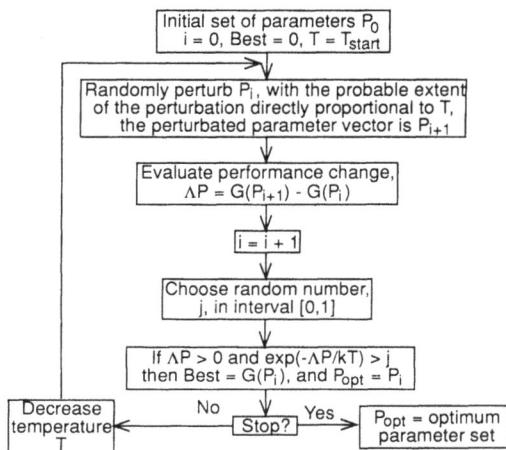

```
                    ┌────────────────────────┐
                    │ Initial set of parameters P₀ │
                    │  i = 0, Best = 0, T = T_start │
                    └────────────────────────┘
```

Figure 5.4 Random-search optimization using simulated annealing.

does gradually cause a finer and finer search of the performance around the present "best" parameters. It has been reported [12] that these techniques perform a sort of adaptive divide-and-conquer strategy. Gross features of the eventual final performance appear at higher temperatures, and fine details develop at lower temperatures. These methods have yet to be applied to yield optimization.

Genetic Algorithms

Genetic algorithms (GAs) are another direct-search optimization strategy. The basis of the procedure is a set of trial parameter sets, sometimes called chromosomes, which are allowed to change or "evolve" towards a set which gives progressively better performance. The key to the genetic optimization is the strategy of change, sometimes likened to survival of the fittest. The idea is that with each change in the parameter population (i.e., each generation of parameters) the performance given by the parameter population improves. This whole process is achieved using a five-step process with the biologically inspired names of representation, evolution, reproduction, breeding and crossover, and mutation.

Representation. Genetic algorithms require the input parameter set to be represented as a string of digits. It is straightforward to map each parameter onto the interval 0 to 1, for instance, and then have each of the n parameters occupy a position in the string of n bounded numbers. The algorithm then manipulates and optimizes this string of numbers as a whole. An individual string of parameters is called an element within the population of parameter strings.

Evolution. Each generation of parameters begins with a performance evaluation of each string in the population. Usually this involves determining the performance G(P) for each representation of P in the population. Each element is then graded as to how well it performed, often times using an error function not unlike those discussed in Section 5.2.1. The GA researchers refer to the error function as the fitness function.

Reproduction. In reproduction, some of the members of the population for the current generation are copied (reproduced) and added to the next-generation population. The number of copies depends on the performance evaluation. Those elements that performed well are copied several times, and those elements with poor performance are not copied at all. The copies, or *offspring,* then make up the next generation. Elements that are not copied are not represented in the next generation. Note that the number of elements in each generation is constant. There are several methods of ranking and reproduction that have been suggested, including *rationing,* where the number of copies is directly related to the element's performance, and *ranking* where the performances are ranked, with the top performers being copied more times than the lower ranked performers [15].

Breeding and Crossover. The previous step, reproduction, produced a population of strings where each evaluated well. Breeding then combines parts of two strings to form two different and new strings. In this way good representations are mixed with poorer representations, with the result eventually being evaluated in the next generation of the algorithm. There are many methods for breeding, the most common being *crossover.* Crossover typically takes two elements, splits them at a random location in the string, and swaps the two parts to create two new strings (Figure 5.5). This is in some sense a controlled statistical exploration of the performance space.

Mutation. The last step in creating a new generation of elements is the random changing of parameters in some of the surviving strings. This comprises a com-

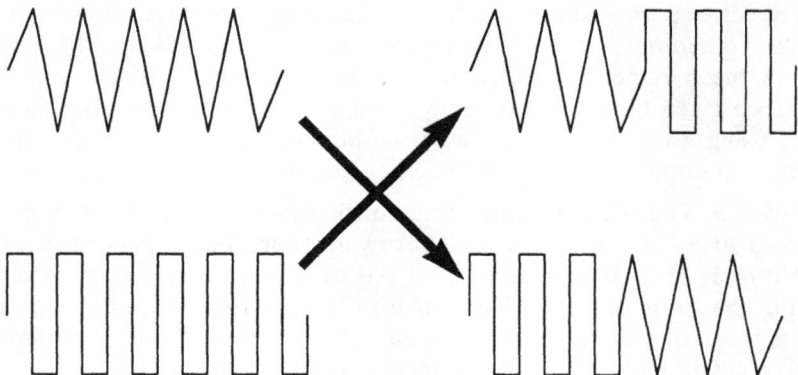

Figure 5.5 Breeding and crossover in the simple genetic algorithm.

pletely random search of the performance space, and can be viewed as the injection of information into the surviving population.

A completed flow diagram of the genetic algorithm as described here is given in Figure 5.6. The application of these techniques requires many tuning parameters, and to our knowledge has not been applied to yield optimization. The authors feel that these techniques will be closely evaluated in the near future and they may prove useful for many complex optimization problems, including discrete-value and tolerance optimization.

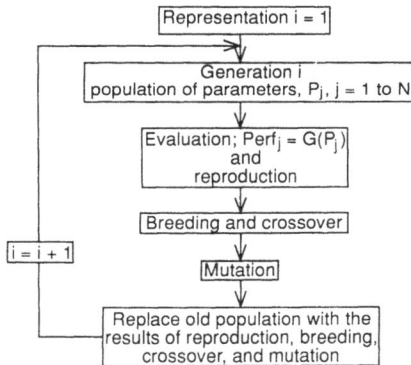

Figure 5.6 Random-search optimization using a genetic algorithm.

5.2.4 Brinksmanship Design

Single-point optimization has been used effectively for unit design. (Just imagine what design would be like without the use of traditional optimization.) Single-point optimization fulfills the task of finding a set of nominal parameter values that minimize the difference between the performance we desire and that which is available from the underlying unit design.

Finding the optimal unit performance is helpful but seldom solves the manufacturability problem because the unit will never be manufactured with the nominal parameter values. This unavoidable parametric uncertainty in manufacturing can result in manufactured performance degradation that may be unacceptable, even for single-point optimized designs. This is called brinksmanship design [8] and was illustrated in Figures 1.12 and 1.13 and is repeated in Figure 5.7 for convenience. It is easy to see that a single-point "optimized" solution P*, is a poor choice for manufacturing because if, due to manufacturing effects, the parameter is slightly larger than P*, the performance degrades below the acceptable level. Thus there is a need to optimize the unit performance, not at a single parameter value, but over the entire range of parameter values that will be encountered

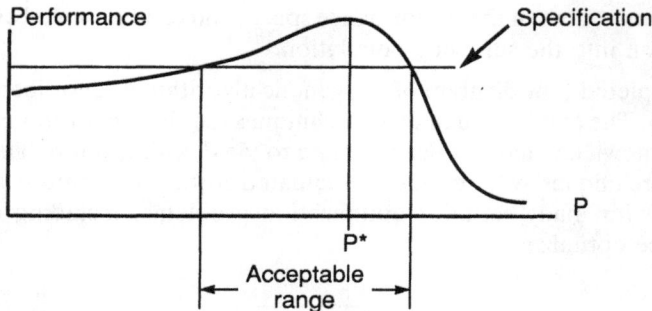

Figure 5.7 Brinksmanship design, a one-dimensional example.

during manufacture. This is the intent of statistical optimization, for which a vast amount of experience from nominal optimization is leveraged.

5.3 STATISTICAL OPTIMIZATION

The previous sections covered the strategy behind single-point optimization—automatically adjust the parameter values in an informed manner until the objective function is minimized (or maximized). Statistical optimization is similar. A statistical optimizer attempts to adjust the unit parameters (now with random variations), until some statistical objective function is minimized (or maximized). But there are inherent differences, both philosophical and numerical, between single-point and statistical optimization. These are most easily illustrated by looking at the yield-optimization problem from both a geometric and mathematical point of view. When approached from a geometric point of view, yield optimization is often referred to as *design centering*.

5.3.1 Design Centering

The yield-optimization problem and some of its properties can be easily visualized in two-dimensional parameter space. As stated in Chapter 3, yield is a measure of the weighted overlap of areas $\mathcal{P}_a$ and $\mathcal{T}$, where the weighting is given by the parameter joint density $f_\mathcal{P}(P)$. Specifically for uniform and independent parameters,

$$\text{Yield} = \frac{\text{area}(\mathcal{T} \cap \mathcal{P}_a)}{\text{area}(\mathcal{T})} \tag{5.9}$$

In this case, for a fixed area of the tolerance region in parameter space, $\mathcal{T}$, the yield is maximized when the area of the intersection of $\mathcal{P}_a$ and $\mathcal{T}$ is maximum.

There are two ways to increase this intersection area: (1) move $\mathcal{T}$, and (2) move and size $\mathcal{P}_a$. The method involving the movement of $\mathcal{T}$ is shown in Figure 5.8.

Making $\mathcal{T}$ smaller (tightening tolerances) will usually increase the yield, however, there is always a practical limit to how small $\mathcal{T}$ can be made. For the current discussion, assume that the parameter variations have been minimized prior to the design-centering process. This implies that the manufacturing process is "under control."

Another often overlooked possibility is to change the size and shape of $\mathcal{P}_a$, the acceptability region. This is done by choosing a different structure or technology for the unit. The effects of structure on circuit yields have only been investigated recently, and then only for simple passive matching circuits [16]. (Actually, another way to size and shape $\mathcal{P}_a$ is to change the unit specifications, S. See, for example, [17].)

In the general case, the definition of a tolerance region becomes, almost literally, blurred. Here, the tolerance region is not so easily distinguished as the evenly weighted rectangle seen in Figure 5.8. Design centering for the general parameter joint probability density function is accomplished by positioning $f_{\mathcal{P}}(P)$ such that it optimally "covers" the acceptability region, $\mathcal{P}_a$. Mathematically this is accomplished as follows. Divide $\mathcal{P}$ into two parts, its mean and its variation about the mean (i.e., $\mathcal{P} = \Delta\mathcal{P} + P_0$), where $P_0 = (p_{10}, p_{20}, \ldots p_{n0}) = E(\mathcal{P})$ and $\Delta\mathcal{P} = \mathcal{P} - P_0$. $E(\mathcal{P})$ is the mean or average value of $\mathcal{P}$, and $E(.)$ is the expectation operation. For this case

$$f_{\mathcal{P}}(P) = f_{\Delta\mathcal{P}}(P)*\delta(P - P_0) \qquad (5.10)$$

where $f_{\Delta\mathcal{P}}(P)$ is the density of the zero-mean random parameter vector P, $\delta(P - P_0)$ is the n-dimensional Dirac delta function, and $*$ represents n-dimensional convolution. Mathematically, the parameter density $f_{\mathcal{P}}(P)$ is equivalent to a density

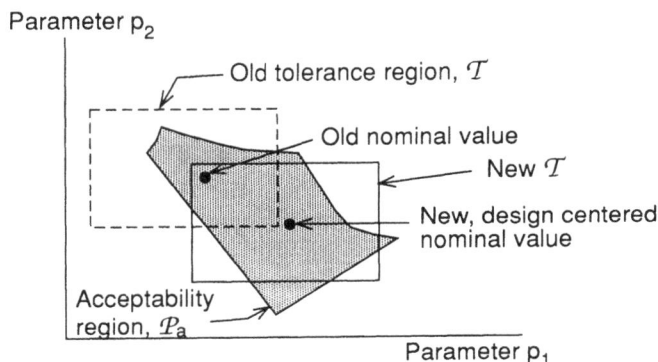

Parameter p_2
Old tolerance region, $\mathcal{T}$
Old nominal value
New $\mathcal{T}$
New, design centered nominal value
Acceptability region, $\mathcal{P}_a$
Parameter p_1

Figure 5.8 Design centering to maximize yield, a two-dimensional example.

centered about zero (i.e., zero mean) displaced to the parameter vector mean value P_0 as illustrated in Figure 5.9.

Allowing the decomposition of $\mathcal{P}$ into $\Delta\mathcal{P} + P_0$ makes $f_\mathcal{P}(P)$ a function of P_0. By moving P_0, we are essentially moving the parameter density in parameter space, without changing its shape. At times we will want to explicitly denote this relation by using the notation, $f_\mathcal{P}(P) = f_\mathcal{P}(P, P_0)$. Using this notation the *fixed tolerance* yield-optimization problem is formulated as

$$\underset{P_0}{\text{maximize}} \ Y(P_0) = \int_{\mathcal{P}_2} f_\mathcal{P}(P, P_0)dP, \tag{5.11}$$

or

$$\underset{P_0}{\text{maximize}} \ Y(P_0) = \int_{-\infty}^{\infty} \text{accept}(P)f_\mathcal{P}(P, P_0)dP \tag{5.12}$$

The integral equations in (5.11) and (5.12) can be recognized as the two forms for the yield expression developed in Chapter 3. The maximization problems as stated in (5.11) or (5.12) are usually referred to as the design-centering problem. The parameter values P_0 are the nominal values set by the design engineer. The variation about the nominal values, described by $f_{\Delta\mathcal{P}}(P)$, is governed by the manufacturing process (and unit environmental conditions, in addition to any other sources of uncertainty). The result of the statistical design process is a set of nominal values which optimally interact with the random manufacturing environment, $f_{\Delta\mathcal{P}}(P)$.

In the context of the general optimization discussion at the beginning of this chapter, the goal for yield optimization is to determine a sequential set of parameter nominal values $\{P_0\}$, such that $Y(P_0^{k+1}) > Y(P_0^k)$. This relation is illustrated over four iterations in Figure 5.10.

The relation between the parameter density, $f_\mathcal{P}(P, P_0)$ and the acceptability region, $\mathcal{P}_a$, is critical to the yield-optimization result. Different parameter densities can result in different optimal nominal values. This is illustrated in Figure 5.11,

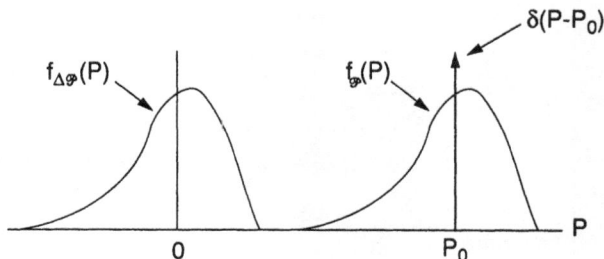

Figure 5.9 The decomposition of $f_\mathcal{P}(P)$ into $f_{\Delta\mathcal{P}}(P) * \delta(P - P_0)$; a one-dimensional example.

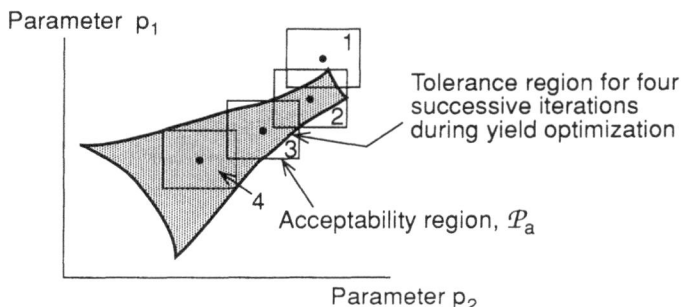

Figure 5.10 The motion of the nominal value during yield optimization for four iterations.

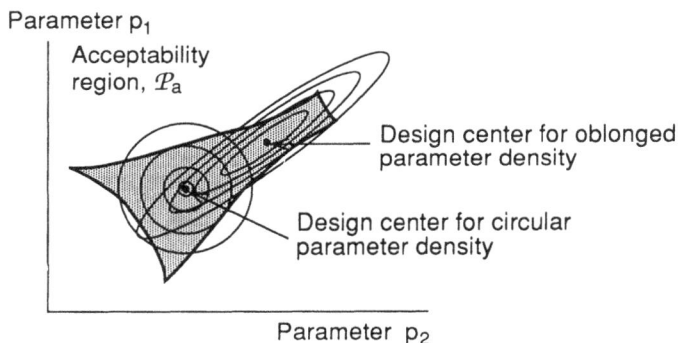

Figure 5.11 The position of the design-centered nominal as a function of the parameter density.

where a two-dimensional circular density, and an oblong density give two completely different design centers for the same acceptability region. Thus the statistical model used in the design centering can have a large effect on the design-centering results.

It is traditional to evaluate a unit by looking at its performance at the parameter value P_0 only. This is the idea behind single-point design and optimization. But the strength behind statistical design and design centering is that the design is evaluated not only at P_0, but at all the points encountered during manufacture, as described and weighted by the parameter density $f_{\mathcal{P}}(P, P_0)$. What we mean by weighted is that points in parameter space that are very likely to occur (i.e., where $f_{\mathcal{P}}(P)$ is large), are given more emphasis in the evaluation than points in parameter space that are not likely to occur. This is in contrast to worst case design where the designer chooses points "far" from P_0, and requires good performance at these points without considering the likelihood that these points will occur during manufacture.

5.3.2 Statistical-Optimization Error Function

In statistical optimization the quantity we wish to minimize (or maximize) is usually some statistical average of the unit performance such as average performance,

$$\bar{G} = \int_{-\infty}^{\infty} G(P)f_{\mathcal{P}}(P)dP \qquad (5.13)$$

performance variance,

$$\sigma_G^2 = \int_{-\infty}^{\infty} (G(P) - \bar{G})^2 f_{\mathcal{P}}(P)dP \qquad (5.14)$$

or yield

$$\text{Yield} = \int_{-\infty}^{\infty} \text{accept}(P)f_{\mathcal{P}}(P)dP \qquad (5.15)$$

Remember that these integrals are over the parameter space $\mathcal{P}$, and as such are n-dimensional integrals.

A comment about two of these error functions is in order. The reader might feel that optimizing average performance will result in acceptable yield. A very simple example will illustrate a problem with average performance optimization. Say we have three manufactured units using two different designs. The performance of the three manufactured units using design 1 is measured at 4, 5, and 9. The performance with design 2 is measured at 5, 5.5, and 6. The average performance of design 1 is 6, while the average performance of design 2 is 5.5. Yet assuming the performance specification requires performance above 4.5, design 1 has a yield of 66.7%, while design 2, the one with the lower average performance, has a yield of 100%. Thus average performance does not tell the whole story. As can be seen, performance variation is also important. Statistical design through yield optimization is a comprehensive way to deal with these problems.

5.3.3 Yield-Optimization Approaches

For most design problems the unit performance, $G(P)$, can be calculated with good precision for any P; $G(P)$ is typically obtained using unit-simulation software incorporating advanced models for both unit components and elements. Equations (5.11) and (5.12) make clear what is required to optimize yield. However, there are two problems that make the yield integral impossible to analytically evaluate and optimize. First, in practice, the parameter joint PDF, $f_{\mathcal{P}}(P)$, is not known explicitly; and second, since insight into $\mathcal{P}_a$ is usually only available in point-wise form through accept(P_k) (see Section 3.5.3), there is a heavy computational burden associated with modeling or exploring $\mathcal{P}_a$.

Primarily for these two reasons, statistical optimization is often treated as a separate subject in optimization. Although statistical optimization has its own inherent difficulties as compared to single-point optimization, the similarities of the two problems should not be ignored. Many different approaches for statistical optimization have been proposed to minimize these two problems. Indeed, the way in which the problems are resolved give rise to two broad categories of methods for the yield-optimization problem: (1) the *deterministic methods,* and (2) techniques based on statistical sampling—*statistical methods.* And while we agree that there is a touch of irony in the notion of "deterministic methods for statistical design" and "statistical methods for statistical design," it is nonetheless the accepted terminology [18, 19]. The following sections examine some of these methods with a focus on yield optimization, although some of the methods could be adapted for use with any of the statistically based objective functions (5.13) (5.14) (5.15).

5.4 DETERMINISTIC METHODS FOR YIELD OPTIMIZATION

Roughly, the deterministic approaches first formulate an approximate analytical model for either $\mathcal{P}_a$ or $f_{\mathcal{P}}(P)$ (or both), and then use optimization theory applied to the closed-form analytical model(s).

5.4.1 Simplical Approximation

As presented in Chapter 3 (Section 3.6.2), the simplical approximation method approximates $\mathcal{P}_a$ using $(n + 1)$-dimensional simplices that are "fit" to $\mathcal{P}_a$ by sampling at points along its boundary. Once $\mathcal{P}_a$ has been approximated, the design center corresponding to maximum yield is determined as the center of the largest hyperellipsoid that can be inscribed within the convex hull defined by the simplices. The design center, P_0^*, illustrated in Figure 5.12 is found using linear programming

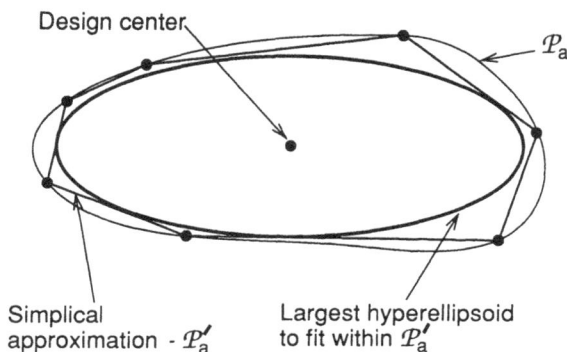

Figure 5.12 Simplical-approximation design centering.

methods. As an additional benefit, useful information about optimal component tolerances can be inferred from the axis lengths of the maximally inscribed hyperellipsoid. Economical Monte Carlo yield estimates are obtained from the simplical approximation (as opposed to the unit simulator).

Aside from the assumptions discussed in Section 3.6.2, the technique assumes that the parameter joint PDF is homothetic [18] (e.g., Gaussian or uniform), where level lines of constant PDF are concentric. The limitations due to the curse of dimensionality as well as the assumptions on $f_{\mathscr{P}}(P)$ and $\mathscr{P}_a$ still hold. The designer should carefully consider the scale of the problems before implementing this method. But nevertheless, the simplical-approximation approach stands out as a classic deterministic method for yield estimation and design centering.

Figure 5.13 shows a yield-optimization summary sheet which gives the essential information about the simplical-approximation approach to yield optimization. The icons used are those described in Figure 3.5. We have included these summary sheets for each yield-optimization method discussed here. The reader can use the sheets as a quick reference to the methods and to compare and contrast the different methods.

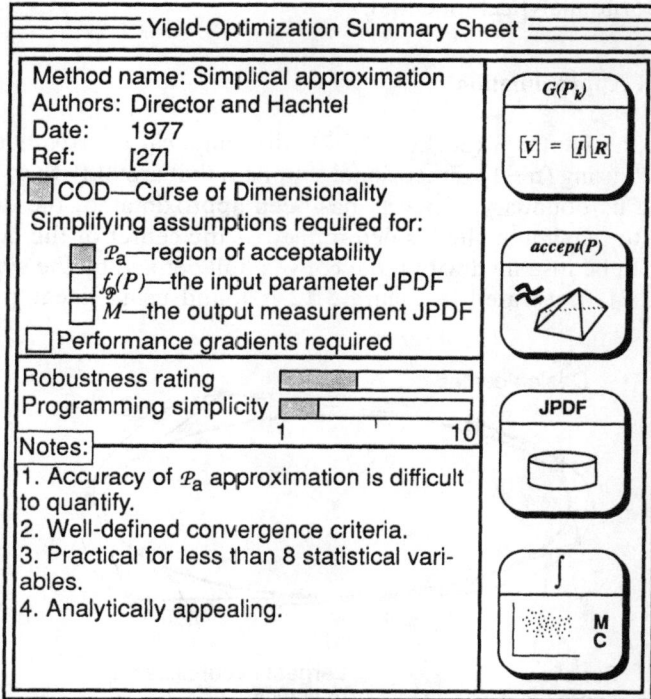

Figure 5.13 Yield-optimization summary sheet using the simplical approximation method.

5.4.2 Multicircuit

Bandler and Chen [4] have proposed a promising method in the multicircuit approach to yield optimization. The method is very intuitive because it has close ties to single-point optimization. Roughly, the multicircuit approach involves the simultaneous application of single-point optimization to not only the nominal circuit, but also to a set of circuits called *auxiliary circuits*. In particular, the method starts by choosing a group of k auxiliary circuits that are related to P_0 according to the parameter joint PDF. These auxiliary circuits form an approximation to $f_{\mathscr{P}}(P)$. Then a multicircuit error function is defined as a weighted sum of the performance errors for each of the k auxiliary circuits. The weights are chosen so the circuits closest to the boundary of $\mathscr{P}_a$ are given more emphasis. This error function is minimized using advanced nonlinear optimization techniques. Minimizing the error function finds P_0 such that the largest number of statistically related auxiliary circuits are within $\mathscr{P}_a$. This essentially maximizes the yield estimate:

$$\hat{Y} = \frac{1}{k}\sum_{i=1}^{k} \text{accept}(P_i)$$

These ideas are illustrated in figure 5.14.

The advantage of this method is the application of powerful, proven gradient-optimization techniques, and the fact that no assumptions on $\mathscr{P}_a$ are necessary. However a different "curse of dimensionality" applies here. In order to obtain a good approximation to $f_{\mathscr{P}}(P)$, k may have to be large—in the hundreds (see Chapter 3). However with this large k, the problem requires the simultaneous optimization of a large number of circuits, a task that may be computationally impractical at this time. To get an idea of the run-time, consider that yield-optimization time relative to the same circuit being single-point optimized is proportional to k. In

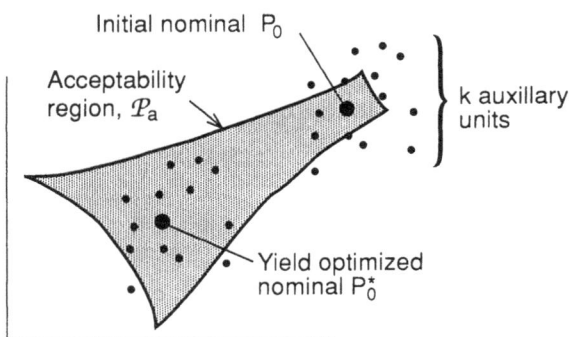

Figure 5.14 Multicircuit approach to yield optimization.

[20], the authors have proposed the use of supercomputers for solving nonlinear microwave-circuit problems with the multicircuit method. Naturally the practicality of the approach improves with smaller numbers of auxiliary circuits, and in fact, empirical evidence with numbers of k on the order of 50 have exhibited good results [20].

As optimization progresses, the algorithm is stopped and then restarted with new selections for the k circuits. This is due mainly to a need to recalibrate the weighting factors of the error functional as the nominal point has moved. Auxiliary circuits near the boundary at the start of optimization are, hopefully, well within $\mathcal{P}_a$ after a few iterations. Therefore, with a new nominal point, it would be advantageous if the driving stimulus for the optimizer stemmed from auxiliary circuits that are on the balance of pass or fail. Restarts are performed until no perceptible change in a Monte Carlo yield estimate is noticed (usually two or three restarts are sufficient).

Recently the multicircuit approach has been combined with the maximally flat quadratic approximation (see Section 3.6.7). By combining an efficient technique for evaluating G(P), Biernacki and Bandler [21] go a long way in alleviating any practicality issues related to the multicircuit approach. Using the fixed pattern of base points, $(2n+1)$ simulations are required around each auxiliary circuit. Then multicircuit optimization is performed on the computationally inexpensive quadratic model. For 200 auxiliary circuits and 20 design parameters, the number of calls to the actual unit simulator is $(2n+1)k$ or 8200 simulations per restart. Conceptually the tradeoff is now between accuracy in the estimate of $\mathcal{P}_a$ and that of $f_{\mathcal{P}}(P)$.

Figure 5.15 shows the multicircuit summary sheet.

5.5 SAMPLING-BASED METHODS FOR YIELD OPTIMIZATION

Deterministic techniques are typified by performing optimization on closed-form analytical approximation(s), and then use Monte Carlo estimation to verify yield improvement. Techniques based on sampling methods use Monte Carlo sampling to obtain yield estimates (and other types of estimated information). The sample data is then used as input to heuristic (perhaps less sophisticated) methods of optimization. Another trademark of the sampling-based methods are their implementation simplicity and "surefiredness."

5.5.1 Statistical Exploration

R.S. Soin and R. Spence [22] developed a *centers-of-gravity* method which uses statistical exploration of $\mathcal{P}_a$. Monte Carlo analysis is performed at the current nominal point P^j. The results of the analysis are used both to estimate yield and to choose new nominal values, which are expected to increase yield. The procedure is repeated until no increase in yield occurs. This heuristic algorithm is based on

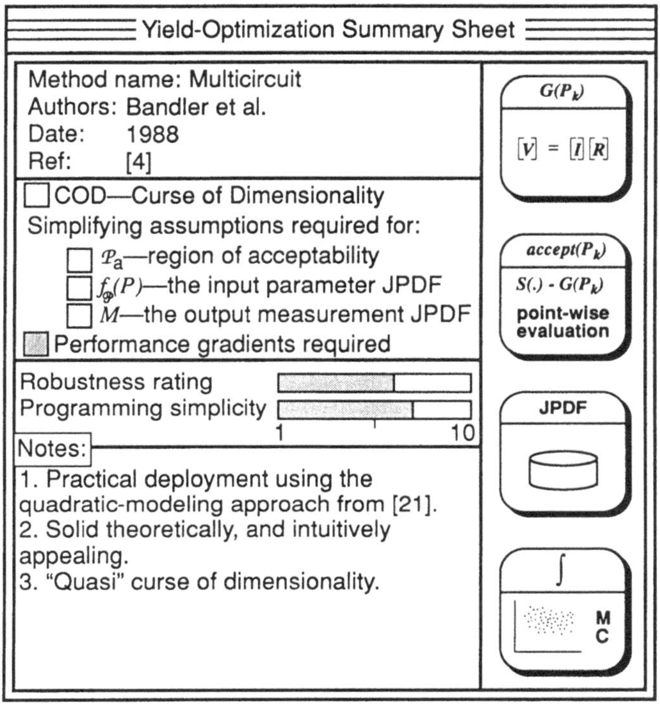

Figure 5.15 Yield-optimization summary sheet using the multicircuit method.

the relative positions, in component space, of the centers of gravity of the pass and fail circuits as identified by the Monte Carlo analysis. A technique reusing samples from one iteration to the next is employed to reduce the number of sample circuits required to be analyzed by each Monte Carlo analysis. Figure 5.16 shows

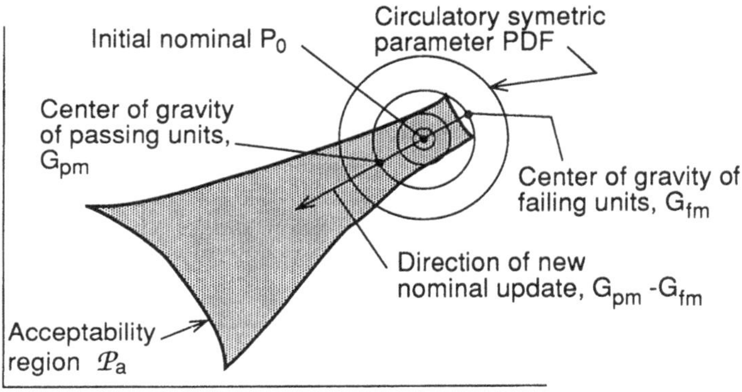

Figure 5.16 Statistical exploration.

the statistical exploration of a two-dimensional acceptability region, while Figure 5.17 shows a flow chart for the centers-of-gravity method.

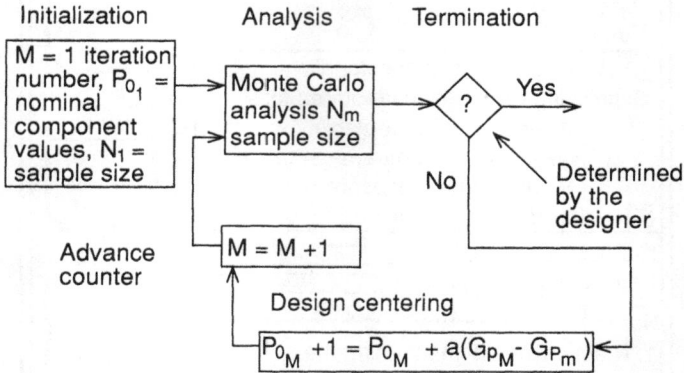

Figure 5.17 Flow chart for centers-of-gravity method for design centering.

5.5.2 Parametric Sampling

Singhal and Pinel [23] developed a statistical method for design centering and tolerancing that does not require Monte Carlo estimation using full circuit simulations for every iteration. The approach borrows the concept from *importance sampling* (see Section 3.3.3) that the yield can be calculated using a parameter joint PDF different from that used to actually interrogate parameter space (i.e., using accept(P_k)).

First, an "inflated" sampling density is used in an initial Monte Carlo exploration that covers the region in parameter space where optimization will hopefully take place. During optimization when a new nominal point, P_0 is chosen, the yield at the new point is recalculated from a weighted sum of the original Monte Carlo simulation results. The weights are the ratio of the actual sampling density value to that of the inflated density value, at the point of the Monte Carlo trial. Figure 5.18 illustrates the weight calculation in parameter space for a single uniform parameter. Yield gradients can also be calculated when the joint parameter PDF is available and differentiable. Singhal and Pinel demonstrated the technique using the multivarate correlated Gaussian joint parameter PDF.

The yield-optimization algorithm uses a quasi-Newton search method based on computed yield gradients. To maintain sufficient accuracy of the yield estimate during optimization, provisions for augmenting the original Monte Carlo database are provided. The only drawback to the method is that it assumes the joint pa-

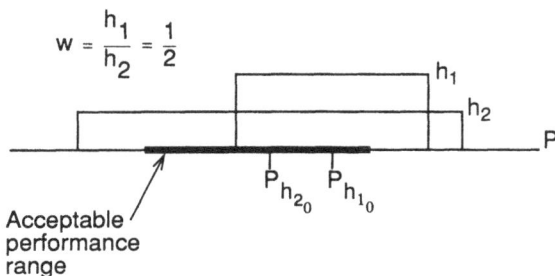

Figure 5.18 Parametric-sampling weight calculation for a single uniform parameter with inflated density h_2, and actual density h_1.

rameter PDF is both available and differentiable—a fairly restrictive assumption when dealing with practical parameters, such as those found in microwave circuits and systems.

Figure 5.19 shows the parametric-sampling summary sheet.

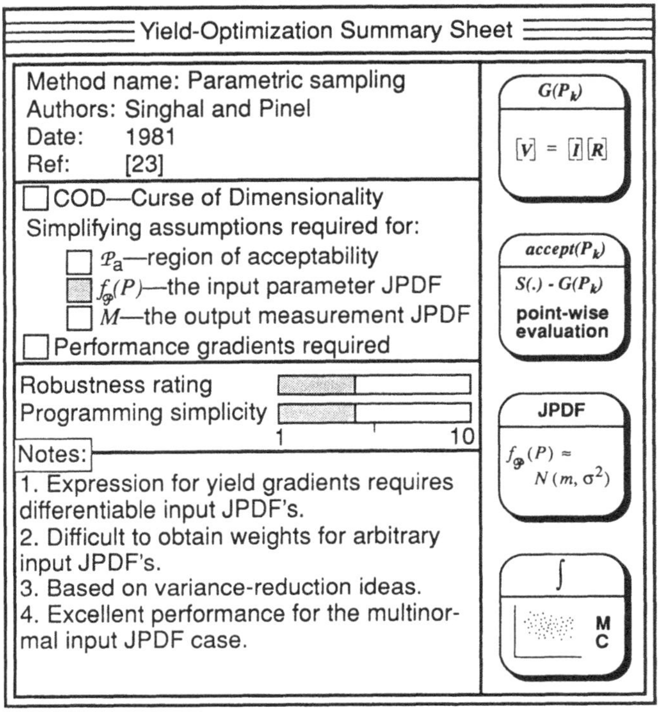

Figure 5.19 Yield-optimization summary sheet using the parametric sampling method.

5.5.3 Radial Exploration

K.S Tahim and R. Spence [24] propose a radial-exploration approach to yield estimation (see Section 3.6.5) and design centering. Recalling from Chapter 3 that $r_{0_j}^+$ and $r_{0_j}^-$ are the normalized Euclidean distances associated with the jth line, the jth *asymmetry vector* can be defined as the vector difference between $r_{0_j}^+$ and $r_{0_j}^-$ (Figure 5.20). The optimization search direction is found by summing the asymmetry vectors for each line used in the radial approximation. The step length along this direction is found in a more heuristic way, where a scaling factor is allowed to increase and decrease according to the radial geometry at the current nominal point.

The observed design-centering capabilities of the radial model are as impressive as the model's yield-estimation capacity. Medium-scale examples were very responsive to design centering where a 57-parameter filter circuit required only three iterations to converge. In fact, all of the examples tested had similar convergence characteristics. Note that for the examples given in the reference, the parameter joint PDFs are assumed uniform and uncorrelated. The number of unit

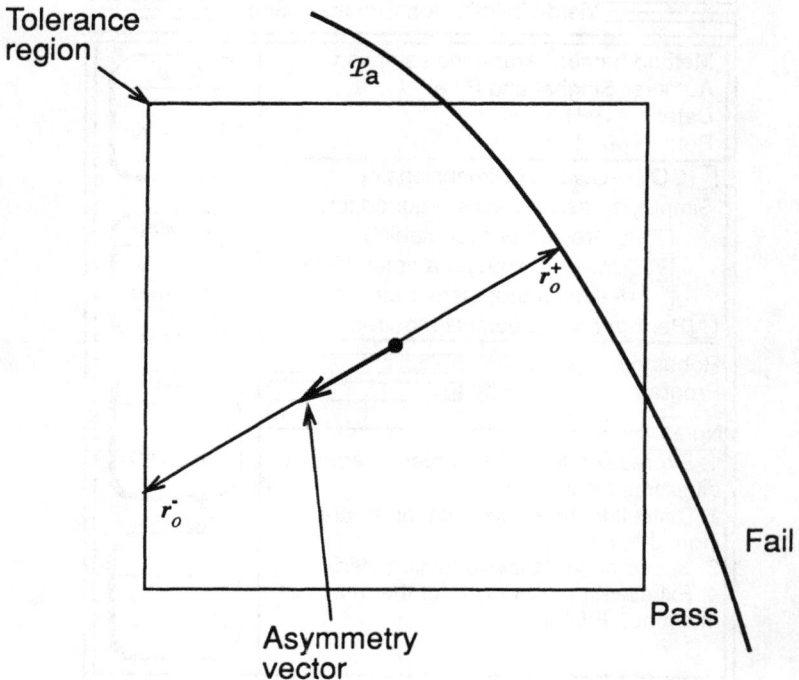

Figure 5.20 Associated direction vectors of the radial-exploration approach to design centering.

simulation calls will depend upon the number of simulations required to find the boundary of $\mathcal{P}_a$ using root-finding techniques. Suppose though that a 50-line model is desired. Further suppose a boundary point identification requires five circuit simulations. This model will require 500 full unit simulations. These numbers are on the order of those found in the multicircuit approach using the quadratic model for G(P). Conceptually, the radial model is a stand-in for accept(P_k), the point-wise approximation for $\mathcal{P}_a$.

Figure 5.21 shows the radial-approximation summary sheet.

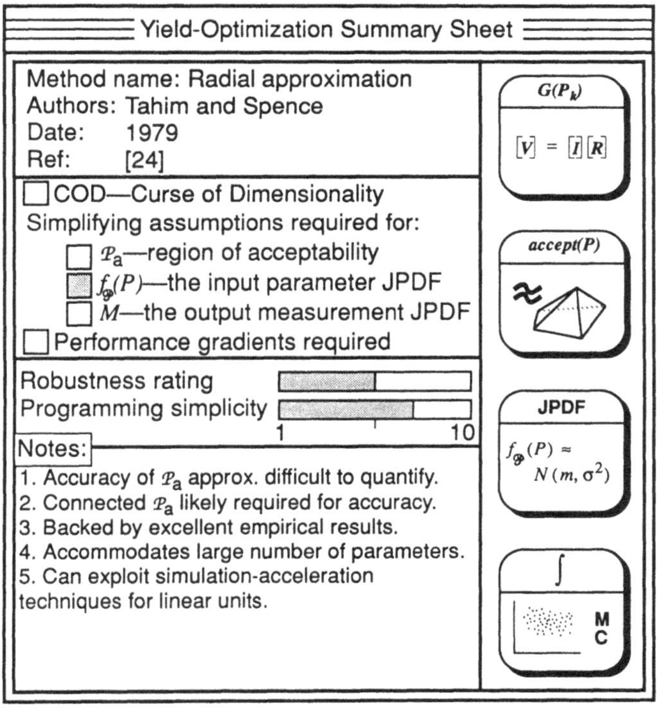

Figure 5.21 Yield-optimization summary sheet using the radial-approximation method.

5.6 YIELD FACTOR HISTOGRAMS

Another useful tool for design centering, tolerancing, and even pinpointing troublesome parameters of a design is available as the *yield factor histogram,* or YFH [25]. The YFH is a graphical display of yield versus parameter value, and it has proven to provide irreplaceable insight into parametric variation of yield versus unit parameters.

A *yield factor* is given by

$$Y(p_{i0}) = \int_{-\infty}^{\infty} \ldots \int_{-\infty}^{\infty} \text{accept}(p_1, p_2, \ldots, p_{i-1}, p_{i+1}, \ldots, p_n)$$
$$f_{\mathscr{P}}(p_1, p_2, \ldots, p_{i-1}, p_i^0, p_{i+1}, \ldots, p_n)$$
$$dp_1, dp_2, \ldots, dp_{i-1}, dp_{i+1}, \ldots, dp_n$$

$Y(p_{i0})$ can be recognized as a Monte Carlo yield estimate with all but the ith parameter varying randomly according to the parameter joint PDF. This relation is mathematically equivalent to letting the probability density function be described as [26]

$$f_{\mathscr{P}}(P) \Leftrightarrow \delta(p_i - p_{i0}) f_{\mathscr{P}}(p_1, p_2, \ldots, p_i^0, \ldots, p_n),$$

where $\delta(p_i - p_{i0})$ is the Dirac delta function.

Note, for example, that if $Y(p_{i0})$ is constant as (p_{i0}) varies over its allowed range of values, then the yield is not sensitive to the ith parameter. Thus the yield factor's slope can be viewed as a sensitivity measure of the yield to the parameter values.

To calculate the yield factor in practice, define an unbiased estimator of $Y(p_{i0})$, $\hat{Y}(p_{i0})$, as

$$\hat{Y}(p_{i0}) = \frac{1}{N} \sum_{i=1}^{N} \text{Accept}(P_i)$$

where P_i is sampled not according to $f_{\mathscr{P}}(P)$ but according to the density $f_{\mathscr{P}}(P)\delta(p_i - p_{i0})$; that is, the ith component of P_i is fixed at p_{i0} and all other parameters are allowed to vary according the density $f_{\mathscr{P}}(P)$.

To further simplify implementation of the yield factor, the values of p_i are divided into m equal-sized regions called bins (m = 9 is often used for convenience). Then to develop an approximation to $Y(p_{i0})$, Monte Carlo sampling is performed where all of the parameters are allowed to vary according to $f_{\mathscr{P}}(P)$, and the m yield factors for each parameter are tallied separately.

Examples of yield factor histograms are given in Figure 5.22 to show what can be gained by such an examination of yield parametric relationships. The YFH in Figure 5.22(a) indicates that the yield is sensitive to the parameter, and that increasing the parameter value may increase the unit yield. That yield *may* increase is due to the fact that the YFHs are smooth functions of the current nominal point, P_0. Nonetheless, YFHs are quite useful in gaining insight into the parametric yield relationships of complex circuits and systems.

The YFH in Figure 5.22(b) indicates that the parameter variation from the lower limit (LL) to the upper limit (UL) is too large. Note that when the parameter

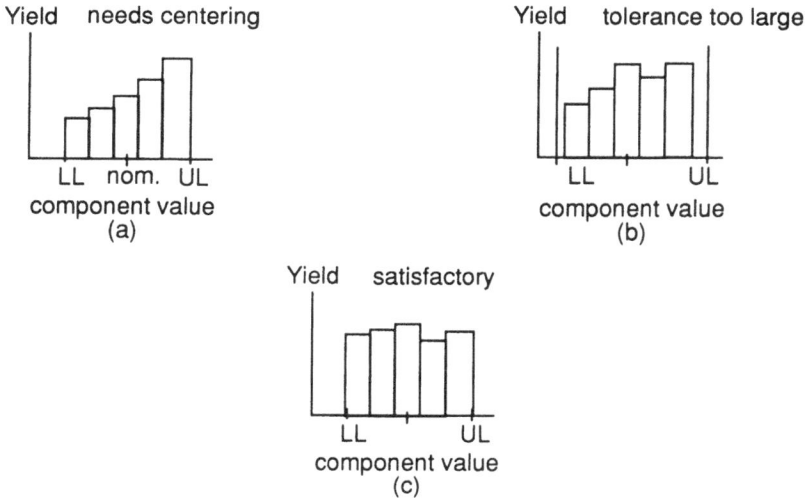

Figure 5.22 Three yield factor histograms showing possible trends of yield versus parameter value: (a) the yield is sensitive to the parameter value and the design needs centering; (b) the lower limit (LL) and the upper limit (UL) of the parameter are too far apart, and the parameter needs tolerancing if possible; (c) the yield is insensitive to the parameter value and is therefore satisfactory.

value is near either limit, the unit yield is zero. In this case, the parameter is a candidate for tolerancing.

Figure 5.22(c) shows that the yield is insensitive to this parameter, and may accept a tolerance increase with little or no effect on the yield. For example, a 10% tolerance resistor might be used instead of a 5% resistor. The design savings realized could be substantial; YFHs give the designer a new set of eyes whereby he can actually see the manufacturing potential of his design.

Figures 5.23, 5.24, and 5.25 depict the relation in two dimensions between the YFHs and the parameter-space descriptions for two cases having uniform and independently distributed parameters. In these figures, the YFHs are presented geometrically along their corresponding axes in parameter space. The value of each bin in the histogram is the ratio of the area of $\mathcal{P}_a$ intersect $\mathcal{T}$ divided by the area of $\mathcal{T}$ defined by the bin (Figure 5.23). Although these figures give an intuitive feel on interpreting YFHs, you should remember that these results are only valid when the parameters are uniform and independently distributed. (Otherwise a suitable weighting factor would be required.)

The following example illustrates the use of a yield factor histogram in the design of a simple electrical circuit. Consider the two-resistor voltage divider shown in Figure 5.26. R_1 and R_2 each have nominal values of 35Ω and are assumed

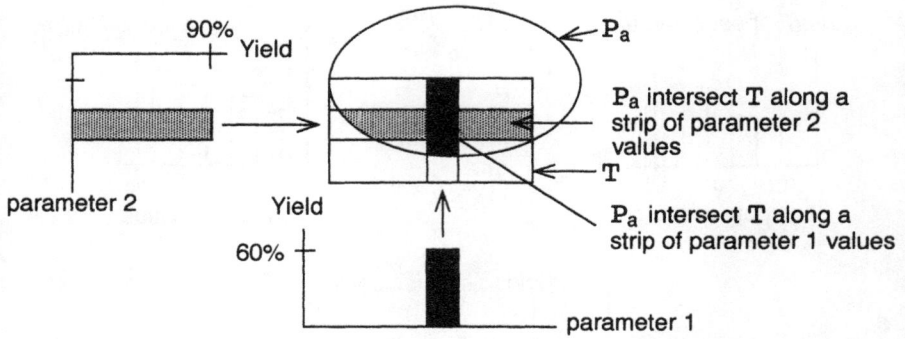

Figure 5.23 Yield factor histograms for two parameters, showing the YFH for a single strip of parameter values for each parameter (i.e., one "bin" of the YFH for each parameter).

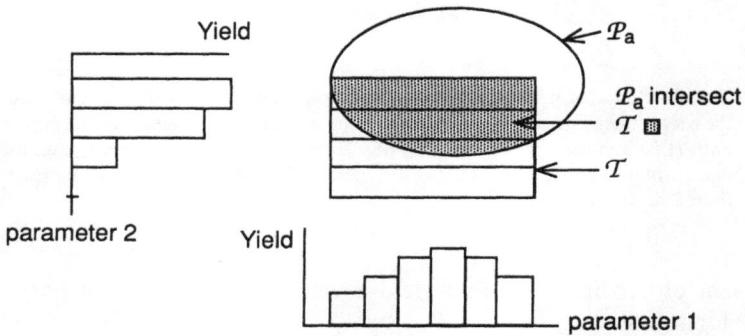

Figure 5.24 Yield factor histogram for a given configuration of $\mathcal{P}_a$ and $\mathcal{T}$ that is independent and has uniformly distributed parameters.

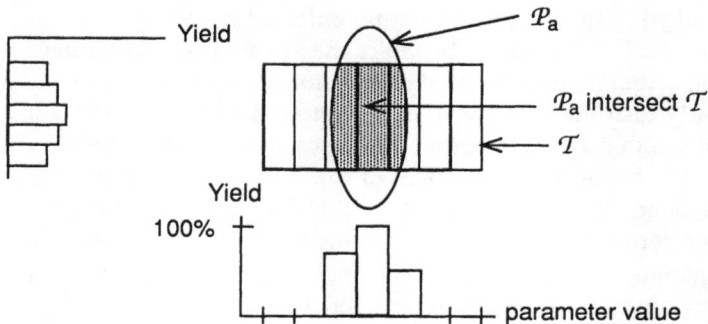

Figure 5.25 Yield factor histogram for a given configuration of $\mathcal{P}_a$ and $\mathcal{T}$ that is independent and has uniformly distributed parameters.

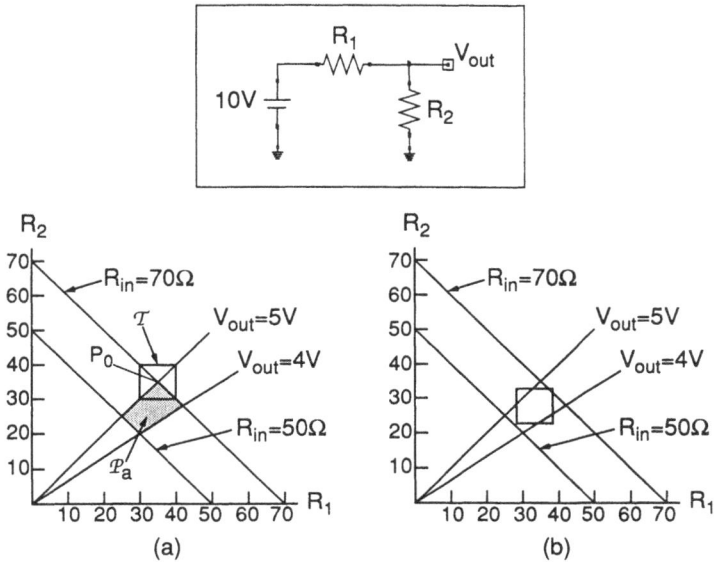

Figure 5.26 A two-parameter voltage divider and its associated geometries in two-dimensional parameter space: (a) before design centering, yield = 25%; (b) after design centering, yield ≅ 83%.

independent and uniformly distributed with tolerances of $\pm 5\Omega$. The parameter space descriptions of $\mathcal{P}_a$ and $\mathcal{T}$ are also depicted in the figure. The specifications for the circuit are $50\Omega < R_{in} < 70\Omega$, and $4V < V_{out} < 5V$.

In Figure 5.26(a), the yield with the nominal at (35,35), for R_1 and R_2 respectively, is 25% because 1/4 of $\mathcal{T}$ is coincident with $\mathcal{P}_a$. Note that a move of the nominal point to (35,25), as shown in Figure 3.18(b), would result in approximately 83% yield because $\mathcal{P}_a$ would have more in common with $\mathcal{T}$.

But this is easy to visualize in two-dimensional parameter space. The YFH view of the situation in Figure 5.26(a) is illustrated in Figure 5.27. Notice the strong gradient toward smaller values of R_2, and a neutral sensitivity with respect to R_1—exactly as expected.

There are several observations to be made about the histograms presented in Figure 5.27. First, the histograms are dependent on the placement of the nominal point P_0. Second, the yield can be obtained for any binned value of a component. For example, for $R_1 = 35\Omega$ and R_2 varying according to its distribution, the yield is approximately 45%. With the YFH you can make component tolerancing decisions. Finally, and perhaps most important, you can use slope information from the YFH for design centering.

Figure 5.28 shows the yield-factor-histogram summary sheet.

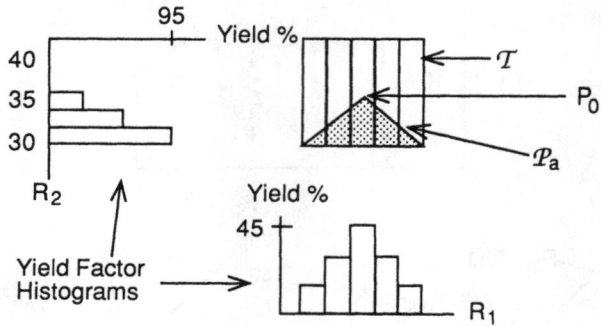

Figure 5.27 Yield factor histograms and their relationship to $\mathcal{P}_a$ and $\mathcal{T}$ for the two-parameter voltage divider.

Figure 5.28 Yield-optimization summary sheet using the yield-factor-histogram method.

5.7 SENSITIVITY REDUCTION

In most units there are variables which vary a great deal, like FET parameters, and variables that are well-controlled, like microstrip line lengths in microwave amplifiers. In this situation, it may be possible to use design centering on the controlled variables to minimize the effects of the uncontrolled variables on the yield. This process is called *sensitivity reduction* and is illustrated in Figure 5.29 for the two-parameter case. Parameter P_1, a line length, has no variation, and P_2, a transconductance, is assumed to vary uniformly. Therefore the tolerance region is a line in component space and the yield is simply the fraction of the line falling within $\mathcal{P}_a$. It is easy to see that a change in the nonvarying component, P_1, can lead to an increase in the yield. This situation illustrates that design centering can improve yield even when some parameters are not controllable.

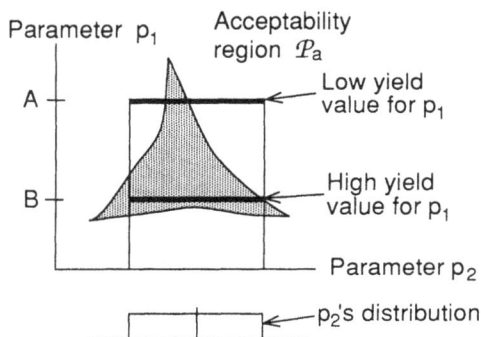

Figure 5.29 Parameter space example showing how careful positioning of a nonvarying parameter can increase yield.

5.8 CONCLUSION

Optimization is becoming a standard part of the design process for sophisticated circuits and systems. Classic optimization generally considers the minimization (or maximization) of a function evaluated at a single set of parameter values. This is called single-point optimization.

Statistical optimization generally involves optimizing some statistical property of the performance function, given a probabilistic model of the parameters. The approaches to both single-point and statistical optimization have been compared and contrasted. There is much similarity between the two optimization problems. Several popular and important statistical-optimization techniques have been detailed.

5.9 IMPORTANT IDEAS FROM CHAPTER 5

Sections 5.1 and 5.2

- Single-point or "nominal" optimization is the optimization of a performance function such that the performance is maximized, (or minimized) at the nominal parameter values.
- Two main classes of optimization techniques are
  1. Gradient methods;
  2. Direct-search methods, which do not use performance gradient information.
- Each optimization technique incorporates a search algorithm and an error-function formulation.
- The generalized least pth error-function formulation is unique in that it allows optimization to continue even after specifications have been met.
- The larger the value for p, the more emphasis is placed on the largest residual.
- Any single-point optimization may lead to a brinksmanship design where nominal performance is optimized, but average performance or yield is poor due to asymmetry in the performance function around the single-point optimum solution.

Section 5.3

- Statistical optimization is the optimization of some statistical property of a performance function such as the average, variance, or yield, or a combination thereof.
- Statistical-design methods allow the designer to optimally trade between the incompatible factors: reliability, cost, and performance.
- Yield improvement can be obtained by changing some or all of unit structure, specifications, parameter tolerances, and parameter nominal values.
- Design centering is the optimal choice of parameter nominal value such that the yield is maximized. For the case of uniformly distributed independent parameters, the design-centering problem has a particularly simple geometric description.
- Design centering for the general parameter joint PDF is accomplished by positioning $f_{\mathscr{P}}(P)$ such that it optimally "covers" the acceptability region, $\mathscr{P}_a$.
- The value of the design center is a function of the parameter joint distribution, and thus is an important input to the statistical-optimization process.

Sections 5.4 and 5.5

- There are two classes of yield optimization: deterministic and statistical. Deterministic approaches generally make an approximation of the parameter statistical models and $\mathscr{P}_a$, and use classic optimization techniques. Statistical approaches are based on Monte Carlo sampling, and use accurate parameter statistical models, but use more heuristic optimization methods.

Section 5.6

* The yield factor histogram (YFH) is a plot of yield versus unit parameter value. It is useful for design centering, component tolerancing, and pinpointing problem parameters within the unit topology.

Section 5.7

* It may be possible to increase the yield by adjusting the value of a highly controllable parameter, so as to minimize the effects of uncontrollable parameters.

REFERENCES

[1] IEEE Spectrum, July 1991.
[2] J. Vlach and K. Singhal, *Computer Methods for Circuit Analysis and Design*, Van Nostrand Reinhold Co., 1983.
[3] J.W. Bandler and C. Charalambous, "Theory of generalized least pth approximation," *IEEE Trans. Circuit Theory*, Vol. 13, pp. 287-289, 1972.
[4] J. W. Bandler, and S.H. Chen, "Circuit Optimization: The State of the Art", *IEEE Transactions on Microwave Theory and Techniques*, Vol 36, No. 2, Feb. 1988, pp. 424-42.
[5] P. E. Gill, W. Murray, and M.H. Wright, Practical Optimization, Academic Press, 1981.
[6] E. Polak, Computational Methods in Optimization—A Unified Approach, Academic Press, 1971.
[7] W.H. Press, B.P. Flannerly, S.A. Teukolsky, W.T. Vetterling, *Numerical Recipes in C*, Cambridge University Press, 1988.
[8] R. Brayton, R. Spence, *Sensitivity and Optimization*, Elsevier, New York, 1980.
[9] T.R. Cuthbert, Jr., Optimization Using Personal Computers With Applications to Electrical Networks, John Wiley & Sons, 1987.
[10] R.A. Rutenbar, "Simulated Annealing Algorithms: An Overview," IEEE Circuits & Devices Magazine, No. 5, 1989, pp. 19–26.
[11] N. Metropolis, A. Rosenbluth, M. Rosenbluth, A. Teller, E. Teller, Journal of Chemical Physics, Vol. 21, No. 1087, 1953.
[12] S. Kirkpatrick, C.D. Gelatt, Jr., M.P. Vecchi, "Optimization by Simulated Annealing," Science, Vol 220, No. 4598, May 1983, pp 671-680.
[13] D. Goldberg, *Genetic Algorithms in Search, Optimization, and Machine Learning*, Addison-Wesley, 1989.
[14] D. Janson and J. Frenzel, "Training Product Unit Neural Networks with Genetic Algorithms," Symposium Digest of the 3rd NASA Symposium in VLSI Design, University of Idaho, Moscow, Idaho, secs. 5.2.1-5.2.8, 1991.
[15] D. Whitley, "Selective Pressure and Ranked Based Allocation," Proceedings of the Third International Conference on Genetic Algorithms, Lawrence Erlbaum Associates, Inc., 1989, pages 116-123.
[16] J. Purviance and D. Monteith, "High-Yield Narrow-Band Matching Structures," IEEE Transactions on Microwave Theory and Techniques, Vol 36, No. 12, Dec 1988.

[17] L. Gefferth, "Specification Sensitivity and its use in System Design," Proc. Inst. Elec. Engr., Vol 129, Pt. G, #4 Aug 1982, pp. 183-185

[18] R.K. Brayton, G.D. Hachtel, and A.L. Sangiovanni-Vincentelli, "A Survey of Optimization Techniques for Integrated Circuit Design," Proc. IEEE, Vol 69, No. 10, pp. 1334-1363, 1981.

[19] R. Spence and R. Soin, *Tolerance Design of Electronic Circuits,* Addison-Wesley Publishing Company, 1988.

[20] J.W. Bandler, Q.J. Zhang, J. Song, and R.M. Biernacki, "Yield Optimization of Nonlinear Circuits with Statistically Characterized Devices," Proc. of the IEEE MTT-S Symposium, Long Beach, CA, June 1989, pp. 649-652.

[21] R.M. Biernacki and J.W. Bandler, "Efficient Quadratic Approximation for Statistical Design," IEEE Transactions on Circuits and Systems, Vol CAS-36, No. 11, Nov. 1989, pp. 1449-1454.

[22] R.S. Soin and R. Spence, "Statistical Exploration Approach to Design Centering," Proc. Inst Elec. Engr., Vol. 127, pt G. pp., 260-269, 1980.

[23] K. Singhal and J. F. Pinel, "Statistical Design Centering and Tolerancing Using Parametric Sampling," IEEE Transactions on Circuits and Systems, CAS-28, No. 7, pp 692-702, 1981.

[24] K.S. Tahim and R. Spence, "A Radial Exploration Approach to Manufacturing Yield Estimation and Design Centering," IEEE Transactions on Circuits and Systems, CAS-26, No. 9, pp. 768-774.

[25] A. McFarland, J. Purviance, et al., "Centering and Tolerancing the Components of Microwave Amplifiers," IEEE 1987 MTT-S International Microwave Symposium.

[26] J. Sarker and J. Purviance, "Yield Sensitivity of AlGaAs High Electron Mobility Transistor," *Int. Jour. of Micro. and Mill.-Wave C.-A. Engr.,* Vol. 2, No. 1, Jan. 1992, pp. 12–27.

[27] S.W. Director and G.D. Hachtel, "The Simplical Approximation Approach to Design Centering," *IEEE Trans. on Circuits and Systems,* CAS-24, pp. 363–372, 1977.

Chapter 6
Statistical Modeling and Validation

"The key to statistical modeling is to create a simulated database which is statistically equivalent to the measured database."

"The designer needs to be able to accurately calculate manufacturing yield, not just improve it."

"Garbage in, garbage out."

6.1 INTRODUCTION

Even with powerful methods for statistical design, if the model for the underlying parameter statistics is inaccurate, then it is likely that design results will be inferior. As an example, the yield integral commonly used in statistical design is given below.

$$\text{Yield} = \int_{\mathcal{P}_a} f_{\mathcal{P}}(P)dP \tag{6.1}$$

Equation (6.1) shows explicitly the importance of the parameter statistics (i.e., the parameter joint density function $f_{\mathcal{P}}(P)$ to the calculation of yield). If there are errors in characterizing the parameter statistics, there are likely to be errors in yield calculation and in yield-optimization results.

    This chapter presents a unified framework for understanding and practicing accurate statistical modeling. The concepts, methods, and modeling ideas presented here can be readily applied to any general system of statistical parameters (passive, active, microwave or otherwise) at any topological level (device, circuit, system, and process), and any operating condition (large and small signal). In this chapter, the statistical-modeling problem is cast in the familiar and highly publicized genre of device modeling, where the modeling procedure typically involves the following steps:

- Characterization;
- Model development and extraction;
- Verification;
- Simulation.

By exploiting the strong parallels in methodology between statistical modeling and device modeling, we hope to provide the reader with a suitable comfort level from which to relate to this critical step in statistical design. A complete treatment of the verification process, as well as characterization of error mechanisms will provide a baseline from which complete and faithful validation of various statistical models can be made. In addition to the proposed framework, we show that not only yield estimates are affected by the accuracy of the statistical model, but also design-centering results.

Over the past two decades there has been significant algorithm advancement in the area of statistical design [1–5]; however, comparatively little attention has been paid to an important aspect: modeling the underlying statistics of random parameters. In fact, recent statistical-modeling works [6–11] show that the mis-application of a given statistical procedure can lead to uncertain results. Thus, part of the intent here is to introduce understandable methods for accurate and reliable statistical modeling.

Based on the authors' experience, there appears to be some confusion about what constitutes a statistical model. Table 6.1 is an attempt at clearing up any misconceptions. The first column contains several "levels" for device characterization, from so-called fundamental process parameters to S-parameters. The second column contains various methods for describing and combining any system of random variables (statistical models). That fact that the parameters for the device models in Table 6.1 are random mandates the use of a statistical model to describe them.

Table 6.1
Examples of Device Models and Statistical Models

Device Model	*Statistical Models*
Fundamental process parameters	Multivariate Gaussian model
Physically-based parameters	Distributed-correlated model
Equivalent circuit element parameters	Regression-based model
S-parameters	Database(empirical) model
	Principle component model

There are presently several competing philosophies on what form a valid and efficient statistical model must take. The physically based FET device models, coupled with multivariate Gaussian statistical models are growing in popularity due to the reasoning that since the device parameters are "close" to the process, a simpler form for the statistical model can be used. Additionally, these models can be scaled, and can track process changes using existing process control data. Also prevalent are the FET equivalent circuit device models combined with empirical database statistical models. The rationale behind their use is both the FET model's simplicity and widespread use, and the fact that there is a "close" relationship between the model parameters and the actual component measurements.

It is not the intent of this chapter to discuss the device models and statistical parameter models in detail, or to recommend one over the other. This chapter primarily proposes a modeling framework which assures that any combination of device model and statistical model is, in fact, accurate in a statistical sense. It is only because the authors are most familiar with database statistical models using either circuit or S parameters that they are used as parameter statistical model examples in this chapter.

Section 6.2 contains a review of pertinent literature. Sections 6.3 through 6.6 are devoted to the development and discussion of three basic aspects of the framework for statistical modeling proposed herein. Since we believe that knowing the limitation of a method is as important as its utility, the limitations of specific methods are discussed. In Section 6.7, we conglomerate the favorable and compatible elements of the previous section into a framework for statistical modeling, which is similar to the familiar methodology used in device modeling.

6.2 SURVEY OF STATISTICAL MODELING

Most of the work in the area of statistical modeling of devices and fabrication processes has been proposed by researchers working with silicon integrated circuits (ICs). Perhaps one of the earliest and most insightful works of the period is Logan [12]. In addition to treating the problems encountered in Monte Carlo analysis of ICs, Logan used statistical techniques to aid in the extraction of performance model parameters. By using the strong correlation between transistor parameters β_N and R_B, Logan was able to separate R_B and R_E from parameters obtained during device measurement; that is, $R = R_B + (\beta_N + 1)R_E$. But as will be shown, the main contribution is the insight made about the so-called average device. As stated in Section 6.6.3, the average device can be a poor representation of a collection of devices.

Rankin [13] surveyed two possible approaches to the characterization of ICs for statistical design. In the first approach, parameters for standard electrical-device models (e.g. SPICE) for the device in question are extracted from measured data. Standard statistical analysis is used to fit distributions to the data, usually assuming normality. Linear (pair-wise) correlation coefficients are used to link the marginal parameter distributions. The second approach uses modeling equations relating the electrical equivalent circuit to the device structure (physics). Application of the statistical procedure known as *factor analysis* is used in an attempt to reduce the number of process parameters required to explain the statistical variations. Like most of the silicon IC statistical-modeling papers, [13] stresses the need to accommodate multilevel statistics; that is, to model the variations across lots, wafers, and chips, and also between matched pairs of elements. No modeled-versus-measured results were given.

Inohira et al. [14] proposed a two-level circuit-oriented statistical model in which pair-wise correlations are modeled both within the device and between devices. In their formulation, Inohira et al. reasoned that two types of correlations in ICs were important: (1) between devices such as two transistors, two resistors, and a transistor and resistor; and (2) between the parameters within a device such as current gain and saturation current for a given transistor. As a special case of the factor-analytic techniques proposed in [13], correlation between devices was modeled using the principal component equation in reverse. The interdevice correlations were modeled using standard regression techniques. Also, simplifying assumptions on the device parameter distribution functions are made by assuming Gaussian or log-normal distributions (between-device distributions are simplified to normal). The measured-versus-modeled verification step used statistical response plots of gain versus frequency for an active filter. While center-frequency variance prediction was good, passband gain variations did not fair so well. We should point out that Inohira et al. considered the use of process-oriented models, but did not use them due to differences between measured and modeled device parameters, along with the need to "tune" the model for a particular fabrication process.

Recently, Strojwas [15] observed that even with powerful computers, general-purpose process simulators [16] or physically based device simulators [17] are computationally expensive and cannot be used in most optimization tasks. To this extent, Strojwas [15] advocates a trade-off between computing efficiency and the "degree of physics" while maintaining accuracy over smaller ranges of process conditions. The statistical-modeling code FABRICS [18,19] uses a mixture of simplified physical models and analytic models to relate input-output quantities for various manufacturing stages. The PROMETHEUS [20] code helps to automate the process of extracting the empirical models and "low-level process-disturbance" statistics. However, as is typically the case [20–24], assumptions on the statistical models (i.e., Gaussian, independent) are imposed to "simplify their characterization." During this period, techniques for efficiently modeling arbitrary joint density functions were not available. In addition to forcing restrictions on the composition of a statistical model, the inability to accommodate arbitrary joint PDFs becomes an issue when testing for *statistical equivalence* between, for example, two data sets—one sampled and the other synthesized.

Statistical equivalence between two data sets establishes that they are from the same joint PDF, within a certain confidence. Spanos and Director [20] were the first (with respect to circuit design) to observe that statistical models that were verified by using only a series of independent univariate statistical tests often give poor simulation results. They called for a *multivariate* test (i.e., a test that will accept or reject the null hypothesis of statistical equivalence for two databases). They reported good results in the ability of their particular methods to model the mean, variance and pair-wise (simple) correlations of various measurable manufacturing quantities.

The microwave-device statistical-modeling effort has a comparatively short history compared to the silicon-IC experience. Purviance and his researchers were among the first to express concern for the state of the art in microwave-device statistical modeling. In [10], the sensitivity of the design center and yield estimate of a FET amplifier were studied as a function of the statistical model. Two different FET statistical models were used: (1) uniform uncorrelated parameter distributions with $\pm 1\sigma$ limits; and (2) marginal distributions and correlations matching those as obtained from the measured data. For the example used, it was reported that the design center was insensitive, while the yield estimate was sensitive to the assumptions made on the FET statistical model. This work also provided evidence that the FET small-signal circuit-model-parameter marginal distributions were not "bell-shaped," and as such required advanced random-variate-generation techniques. It is important to note (as mentioned in [10]) that the experiment used FET statistical models which were not statistically validated. In short, the statistics of the simulated S-parameters were not compared to those of the measured S-parameters.

Meehan and Collins [11] extended the work of [10] and provided the first evidence that statistical models based on FET parameter marginal PDFs and simple correlations were not capable of correctly capturing even the low-order S-parameter statistics (i.e., mean and variance) as estimated from the measured data. Meehan and Collins [11] used the intrinsic FET model with 1-GHz S-parameter data so that the mapping from S to FET parameters is unique, thereby eliminating any effect due to parameter extraction. Using the example from Purviance [10], Meehan and Collins [11] demonstrated that both the yield estimate and design center were sensitive to the statistical model used (see Section 6.6.5).

The results from [11] were later presented in [9], along with a proposal for a viable and valid statistical model called the *truth model*. The truth model is a database model which acts as an empirical discrete density function based on measured samples. While the truth model given in [9] utilized FET equivalent circuit model parameters, the database concept is equally applicable to any statistical characterization.

In [7], scaling of a 200-μm FET truth model (obtained from Watkins-Johnson production line PCMs) to accommodate 300-μm devices demonstrated certain practical aspects of the database model, by accurately predicting the statistical response of a production GaAs MMIC amplifier. (See the Case Studies in Chapter 7.)

Anholt [6] investigated how extraction of FET model parameters affects the ability of database models to predict the low-order statistics of the measured S-parameters. All extraction methods tested indicated predictive powers similar to those discussed above. Anholt compared means and standard deviations which, for this case, provides only necessary (not sufficient) proof of statistical equivalence.

Reminiscent of Inohira et al. [14], factor analytic methods in reverse were applied to microwave FET S-parameter data by Purviance, Petzold, and Potratz [25]. While agreement between measured and modeled S-parameter correlations was demonstrated, the S-parameter marginal distribution agreement was only fair. Purviance, Petzold, and Potratz's [25] main aim was to exploit the independence of the factors in an effort to simplify the process of accurate random parameter simulation.

Physics-based microwave-device models have also been reported in the statistical-modeling literature. Their popularity in this arena stems from a desire to find the device model with parameters having "simple" statistical relationships. This is supported by the view of many authors that because physical models relate to geometrical and process parameters, their statistical distributions will be "closer" to Gaussian and "less" correlated [26]. We note that the statistical assertions must be accompanied by adequate statistical testing and experimentation. Usually S-parameter averages and standard deviations form the basis for validation, but these are only necessary (not sufficient) criteria for delineating between measured and modeled joint densities. Furthermore the results reported are presently inconclusive. For example, in [27], errors in g_m mean and standard deviation measure 17% and 500% respectively. Authors [28] often attribute disparities between measured and simulated S-parameters, mean and standard deviation, to "a deficiency of the model in matching the measurements of the individual devices." While the statistical impact of device modeling errors should be scrutinized, a likely cause for the reported disparity is certain assumptions made in the overall statistical-verification experiment. Specifically, many authors assume

$$E\{S\} = G^{-1}(E\{G(S)\}) \qquad (6.2)$$

where S is the random vector of S-parameters, $E\{.\}$ is the expectation operator, $G(.)$ transforms S-parameters to physical model parameters, and $G^{-1}(.)$ transforms physical model parameters to S-parameters. Equation (6.2) is only true for the case where $G(.)$ and $G^{-1}(.)$ are linear. When (6.2) is found invalid, it may not be a modeling problem, but rather a property of the nonlinear transformation of random variables.

Additionally, (and not unlike the silicon-IC physics-based models) the increased complexity of physically based models is usually accompanied by longer simulation times [29]. This does not suggest that the physical models are poor. In fact, such models are extremely attractive because they offer an economic solution to the statistical characterization of any and all devices produced by a foundry. Also, physical models may require fewer data be taken for validation. Since data collection is costly, it needs to be considered when comparing modeling techniques. Finally, we should point out that the statistical modeling framework proposed here is fully capable of accommodating and validating such models.

Notice a key theme in all of the works cited above: the need both to describe and to generate outcomes for a multivariate joint density function. However, due to a lack of applicable multivariate methods, simplifying (univariate) assumptions on both the statistical models and the verification techniques have been used. But in the general case, a given set of univariate marginal densities and simple correlations can describe any one of an infinite number of joint PDFs. The modeling framework described here uses the full available statistical description for validation, rather than just marginal densities and pair-wise correlations. The framework is novel in that it imposes no restrictions on the modeling of, or delineation between, any joint PDF. The following section discusses the elements of statistical modeling, leading up to formal introduction of the framework.

6.3 ELEMENTS OF STATISTICAL MODELING

Since the steps involved in device-performance modeling (device modeling) are strikingly similar to those of statistical modeling, we first examine the familiar process of device modeling. The goal in device-performance modeling is to develop a mathematical model which accurately predicts the actual measured performance of the device. While this sounds simple on the surface, it is usually a punctilious process (Figure 6.1a).

Single-device-performance characterization begins with an assessment of which terminals and operating conditions of the device need to be measured. Measurement data are then taken. The model development and extraction step determines (1) the functional relationships which best fit the measured data, and (2) if sufficient data have been obtained. The extraction and verification processes are rolled into one step. Extraction is essentially the procedure of determining the values of the model parameters that minimize discrepancies between the measured data and modeled results. Nonlinear optimization algorithms are usually employed to automate this process, although interactive hand tuning can be beneficial in the early stages of extraction. The verification stage of the modeling process involves the quantitative analysis of the errors between the modeled results and the measured data. If errors are too large, a refinement of the model is necessary via additional characterization iterations. If errors are acceptable, then the new device performance model can be put to use in a circuit-design CAD package.

The device-statistical-modeling process is conceptually no different than the device-performance-modeling process (see Figure 6.1(b)). While the steps in the modeling processes are the same, the characterization, modeling, and verification

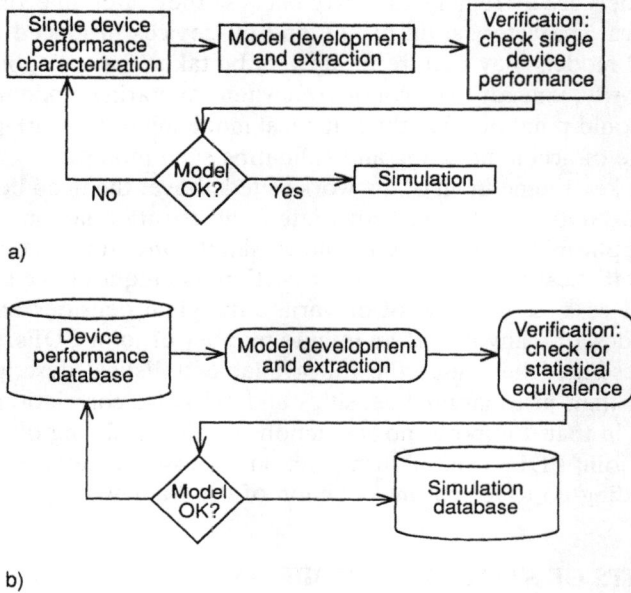

Figure 6.1 The device-modeling process: (a) single-point performance modeling; and (b) statistical modeling.

formulations are different because they are statistical in nature. For example, instead of measuring just one device, statistical device characterization requires many measurements to obtain a statistical database. Therefore, the goal in device statistical modeling is to obtain (by proper formulation and adjustment of model parameters), statistical equivalence between the measurement database and a database simulated from the statistical model. If we are to say that we have a valid statistical model, then statistical equivalence must exist between the measured data and the simulation database generated using the statistical model. This assumes that the devices measured are the ones to be used in the circuit, and the measurement setup does not significantly alter the measurements. Note again that this is not unlike device-performance modeling because model results must agree with the measured data, and measurement considerations are very important.

In the following three subsections we present details and examples of our statistical modeling framework shown in Figure 6.1(b). Section 6.4 presents characterization, Section 6.5 presents verification, and Section 6.6 presents model development and extraction.

6.4 STATISTICAL CHARACTERIZATION

Characterization involves two steps: (1) determining what to measure, and (2) taking measurements. Determining what to measure appears to be the same for both performance and statistical modeling problems. For instance, S-parameters over frequency are usually used to characterize the performance of n-ports.

However, there is a big difference between measurements for performance modeling and statistical modeling. For performance modeling, a single device is measured and the model is fit to the measurement. But the purpose of the statistical model is to characterize the "randomness" of device parameters, and therefore involves the measurement of many devices. By the nature of the problem, a larger data-collection effort is necessary to accurately characterize the device parameter statistics. (Note we could just as well be referring to process or system parameters.) The collection of all the measured data is called the measurement database or population sample. Statistical-estimation methods should be used to ensure that the process of obtaining samples (as well as the quantity of samples taken) is "representative" of the underlying population. Inferences made concerning a population (by use of samples drawn from it), together with indications of the accuracy of such inferences (using probability theory), form the basis of well-established *statistical-inference* methods [30].

Given that the accuracy of the population estimate increases with each additional sample, some form of cost analysis will undoubtedly enter into the characterization process. Subsequently, the sampled database accuracy may influence the requirements for model accuracy. (There is little benefit to accurately modeling a poor representation of the underlying population.)

Because the statistical model will likely be used to calculate yield, it is informative to relate the accuracy of the yield estimate to the size of the sample population [1, 31, 32]. For instance, for an arbitrary distribution and a yield of 50% (a worst case situation), a 95% confidence level gives an error in the estimate of $\pm 3\%$ for 1067 sample data points. For 96 and 266 sample data points, the errors in the yield estimate are $\pm 10\%$ and $\pm 6\%$ respectively (for a 95% confidence level). (The statistical interpolation technique proposed by Campbell [33] may be of use in reducing the sampling requirements [see Section 6.6.6].)

We conclude that the result of the characterization process is a database of measured samples, made over a sufficient and cost-effective sampling of device parameters. Then, the key to statistical modeling is to create, using the statistical model, a simulated database statistically equivalent to the measurement database. Metrics have been developed to verify the statistical equivalence of two databases.

6.5 VERIFICATION

The verification step in statistical modeling measures the "statistical accuracy" of a given combination of device model and statistical model. Only a small quantity of literature addresses the accuracy of the statistical models used for statistical design and yield estimation ([6–10, 34] for example). The problem in verification is determining what statistical equivalence test to perform on the measured and modeled data samples. Two databases are statistically equivalent, if they are samples from the same joint probability density function, with a high degree of confidence. In [20] Spanos and Director point out that a series of independent, univariate statistical tests fail to provide adequate proof of statistical equivalence between joint PDFs.

Notice that univariate tests (Kolmogorov-Smirnov, mean, standard deviation, kurtosis, skewness, Chi-squared, etc.) are necessary but not sufficient indicators of statistical equivalence. Meehan and Collins [11] provided evidence of this by showing that two data sets of FET equivalent circuit parameters, one measured and the other synthesized, and having identical marginal distributions and pairwise correlations, produced sets of FET S-parameters with different univariate statistics.

The present-day literature makes a compelling case that multivariate statistical tests should be used in the overall model verification step.

6.5.1 Tests for Multivariate Statistical Equivalence

For one-dimensional data, there are several tests which can determine statistical equivalence between two sample databases, such as the Chi-squared and Kolmogorov-Smirnov (K-S) tests. However, these tests either do not scale directly to higher dimensions or else they do not have adequate "power" in higher dimensions. Two promising solutions to the problem of determining the multivariate statistical equivalence between two databases are those presented in Friedman [35], (generalized K-S test) which uses minimal spanning trees to generalize one-dimensional tests to higher dimensions, and a new approach based on nearest-neighbor-type coincidences presented in Schilling [36] and Henze [37]. (See also [41].)

6.5.2 The Generalized Kolmogorov-Smirnov Test

The Kolmogorov-Smirnov test compares two sets of samples by measuring the maximum deviation between the cumulative distributions of the samples. In one dimension, the K-S test works by ordering the combined samples and measuring the percentage of samples of the opposite type less than each sample. The significant statistic is the maximum difference between the percentage of samples from the opposite set which are less than a given sample.

In order to make the K-S test work in higher dimensions, it is necessary to define an ordering on samples which is meaningful in higher dimensions. In Friedman [35] minimal spanning trees (MSTs) are traversed in a "height-directed preorder" pattern. A spanning tree is a noncyclic graph containing all points in the space. A minimal spanning tree is a spanning tree where the edges of the tree are weighted by the distance between points, and the sum of the weights is a minimum. The traversal of a MST is a recursive algorithm: visit the root of the tree, then traverse the subtrees of the root in order of least to greatest maximum depth of the subtree. The traversal of the minimal spanning tree defines an order that can then be used by the one-dimensional KS test. Algorithms for constructing minimal spanning trees can be found in most common algorithm books.

6.5.3 Nearest Neighbor Test

A new approach to solving the two-sample problem for higher dimensions is presented in Schilling [36] and Henze [37], and is based on the number of nearest-neighbor-type coincidences. The basic idea is to find the k nearest neighbors in the combined space of the two data sets according to a given distance measure. A statistic is then computed by the following formula [36]:

$$T_{k,n} = \frac{1}{nk} \sum_{i=1}^{n} \sum_{r=1}^{k} \ell_i(r) \qquad (6.3)$$

where n is the number of samples, k is the number of nearest neighbors, and ℓ_i is unity if the rth nearest neighbor is of the same type (sample) and zero otherwise. In Schilling [36], the asymptotic distribution of the statistic is found to be Gaussian for a Euclidean distance measure. In Henze [37], the asymptotic distribution of the statistic is given for any distance measure. Henze's asymptotic distribution is somewhat more complicated than Schilling's, and will not be discussed here.

6.5.4 Multivariate Verification of FET Data

In this example of statistical verification, we use the measured GaAs FET data from [9]. Using a careful synthesis process, another FET equivalent circuit parameter database is produced which closely matches the univariate statistics. It is significant to note that the synthesized model makes no simplifying assumptions about the distributions or correlations—they are very similar to those measured. Table 6.2 shows the results from the univariate K-S test which indicate that the univariate marginal distributions for the measured and synthesized data are statistically equivalent.

Table 6.2
Univariate K-S Test on Measured versus Synthesized FET Parameter Data Sets

R_i	→ Var 0 - KS statistic = 0.015909 with confidence = 1.000000
R_{ds}	→ Var 1 - KS statistic = 0.011818 with confidence = 1.000000
g_m	→ Var 2 - KS statistic = 0.011818 with confidence = 1.000000
C_{ds}	→ Var 3 - KS statistic = 0.010909 with confidence = 1.000000
C_{gs}	→ Var 4 - KS statistic = 0.012273 with confidence = 1.000000
C_{dg}	→ Var 5 - KS statistic = 0.013182 with confidence = 1.000000
t	→ Var 6 - KS statistic = 0.013636 with confidence = 1.000000
G_{dg}	→ Var 7 - KS statistic = 0.014091 with confidence = 1.000000

Table 6.3 shows the linear pair-wise correlation matrix for measured and synthesized data, again with similar results. However, when the multivariate test statistic from (6.3) is used, we obtain a confidence level (consult [36] for computational details regarding μ_i and σ_i):

$$\left| \operatorname{erf}\left(\frac{\sqrt{nk}}{\sqrt{2}} \frac{(T_k - u_k)}{\sigma_k} \right) \Big|_{k=8} \right| = 0.99983$$

Table 6.3
Pair-Wise Correlations on Measured and Synthesized FET Parameter Data Sets.

The correlation matrix for 88 measured devices

	0	1	2	3	4	5	6	7
0	1.000							
1	0.487	1.000						
2	0.068	0.597	1.000					
3	−0.851	−0.454	−0.279	1.000				
4	0.259	0.601	0.820	−0.485	1.000			
5	−0.067	−0.692	−0.165	−0.051	−0.144	1.000		
6	−0.062	0.233	0.080	0.048	0.446	−0.288	1.000	
7	0.677	0.527	0.130	−0.629	0.082	−0.350	−0.403	1.000

The correlation matrix for 200 synthesized devices

	0	1	2	3	4	5	6	7
0	1.000							
1	0.421	1.000						
2	0.002	0.556	1.000					
3	−0.809	−0.361	−0.185	1.000				
4	0.266	0.555	0.750	−0.450	1.000			
5	−0.065	−0.690	−0.178	−0.070	−0.067	1.000		
6	−0.104	0.204	0.020	0.145	0.363	−0.274	1.000	
7	0.637	0.481	0.118	−0.621	0.086	−0.320	−0.373	1.000

This indicates that we reject the hypothesis, H_o, that the measured and synthesized data are statistically equivalent, with a 99.98% confidence level. This explains the results given in [9] and [11], where univariate characteristics of the measured S-parameter data did not compare well with the synthesized S-parameter characteristics, generated from the FET parameter data as given above. If we examine scatter plots of the FET data, we can appreciate why the multivariate test rejects H_o. Figure 6.2 shows R_i versus τ for (a) the measured data set, and (b) the synthesized data set. It is easy to recognize areas where combinatorial discrepancies are prevalent, yet there is excellent agreement between the distributions and correlations involved. This is the main theme: ensembles of univariate test statistics are not suitable for full characterization of multivariate joint probability density functions.

Finally, the multivariate K-S test was applied to the synthesized FET data set—"split in half." Thus we would expect that this test would indicate that the data sets are from the same population. This test indicated statistical equivalence with an 88% level of confidence.

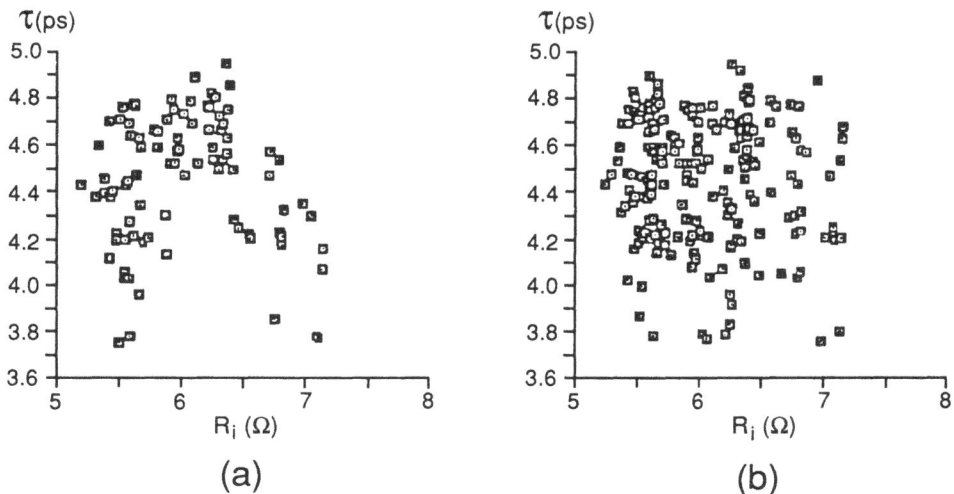

Figure 6.2 Scatter plot of τ versus R_i for (a) measured, and (b) synthesized FET parameter data sets.

6.5.5 Summary of Multivariate Statistical Verification

We have examined techniques for objectively testing multivariate statistical device data for statistical equivalence. These tests are straightforward to implement and provide high power—distinguishing between joint probability density functions from sampled data with calculable accuracy. By distinguishing between two GaAs

FET databases having nearly identical univariate statistics (i.e., marginal distributions, means, standard deviations, kurtosis, skewness, and linear correlations), we are able to gauge the practical utility of these verification methods.

6.6 STATISTICAL-MODEL DEVELOPMENT AND EXTRACTION

The statistical-model-development step involves determining statistical relations within the measurement database and developing analytic expressions which can reproduce those relations in simulated data. Many forms of statistical models have been proposed, including those based on low-order statistics such as:

1. Factor-analytic methods [14, 21, 24, 25];
2. Multivariate Gaussian with linear, pair-wise correlation [9, 29];
3. Univariate marginal densities with or without simple correlation [9];
4. Combined multivariate Gaussian and discrete density functions [31].

The goal of the extraction process is to determine the parameter values of the statistical model that provide statistical equivalence between the databases of simulated and measured data. Because univariate statistical moments are easy to extract from a database, they have been used extensively as the basis for statistical models, as illustrated by the four modeling forms listed above. But these low-order-moment-based models are usually not sufficient to characterize multivariate data, as will be shown in the next section. Two recent database models that are "distribution-free" (e.g., no assumptions made on the form of the device parameter joint PDF) are the truth model [9], and the closely related statistical interpolation model [33]. First though, we provide motivation by examining a hypothetical example using low-order moments to model FET parameter database statistics.

6.6.1 Design Scenario Using the "Average" Device

Suppose we are tasked with the design of a small-signal amplifier to be used as a general purpose gain block (i.e., high gain, good i/o match, and moderate noise performance). Since recent wafer test data is available on the chosen device, we decide to characterize the average device. We send our trustworthy technician into the lab and he returns sometime later with a data disk having 88 records. Each record contains the extracted FET parameters for a particular device sample (with a bias of 0.5 Idss over the frequency range from 1 to 26 GHz). The sampled values for C_{ds} are printed in Table 6.4. Taking the average of the values shown in Table 6.4, as well as for other intrinsic model parameters, yields the "average" device as shown in Table 6.5.

After an initial foundry run of the amplifier design, we find that the predicted amplifier response based on the "average" FET parameters only scarcely matches the measured amplifier responses. What went wrong? Did the process shift? Will another design iteration solve the problem? As microwave-circuit designers (not

Table 6.4
FETFITTER® Extracted C_{ds}(nF) Values From 88 Measured FETs

0.1058	0.1041	0.1044	0.1020	0.1045	0.1029	0.1049	0.1039
0.1029	0.1004	0.1008	0.1021	0.1032	0.1021	0.1019	0.1028
0.1125	0.1141	0.1129	0.1103	0.1119	0.1123	0.1133	0.1117
0.1114	0.1102	0.1120	0.1114	0.1145	0.1149	0.1159	0.1148
0.1115	0.1123	0.1131	0.1129	0.1125	0.1116	0.1144	0.1151
0.1144	0.1159	0.1179	0.1162	0.1192	0.1179	0.1173	0.1153
0.1144	0.1161	0.1164	0.1154	0.1162	0.1159	0.1148	0.1147
0.1135	0.1130	0.1111	0.1120	0.1125	0.1133	0.1141	0.1162
0.1159	0.1131	0.1131	0.1144	0.1151	0.1120	0.1165	0.1150
0.1020	0.1030	0.1045	0.1031	0.1033	0.1026	0.1058	0.1044
0.1068	0.1057	0.1047	0.1034	0.1064	0.1048	0.1080	0.1063

Table 6.5
The "Average" FET

R_{in} = 6.09Ω	R_{ds} = 465.01Ω	C_{ds} = 110 fF	C_{gs} = 433 fF
C_{gd} = 32 fF	G_m = 33 mS	G_{dg} = 91 mS	t = 4.445 ps

statisticians) we must understand the problems posed by this example. The basic problem here is that the FET statistics were not adequately modeled by the average values of the FET parameters. We will see this more clearly in the next sections.

6.6.2 Moments

A number of statistical indicators called *moments* may be useful to generally describe a collection of measured data (like our FET measurement database). Self moments are calculated as sums of integer powers of the individual outcomes of the sample. (A number of these moments were presented in Chapter 2, and appear in Table 2.1.) For example, the mean value of the C_{ds} sample from Table 6.4 is estimated as

$$\hat{\overline{C_{ds}}} = \frac{1}{N}\sum_{i=1}^{N} C_{dsi} \qquad (6.4)$$

where N = 88 is the size of the sample. $\hat{\overline{C_{ds}}}$ is an unbiased estimate of the true mean given by

$$\int_{-\infty}^{\infty} C_{ds} f_{C_{ds}}(C_{ds}) dC_{ds} \qquad (6.5)$$

Of the other moments, most notable are the first through fourth: the mean, variance, skewness, and kurtosis, respectively. The variance is used to denote the amount of spread or dispersion from the mean. The skewness indicates the degree of asymmetry of a distribution, while the kurtosis indicates the degree of "peakedness" of a distribution.

Note that these moments usually make an assumption about the shape of the sample distribution. For example, what do skewness and kurtosis indicate in the context of a bimodal distribution? These measures are really intended for distributions "not too different" from Gaussian.

Also notice that the sample moments transform all of the statistical properties representing a data sample and compress them into a single number. Because one cannot usually obtain something for nothing, the ease with which moments are calculated is offset by the "completeness" of the information they provide. To use moments blindly, without regard for the underlying probability distribution of the sample is an invitation for surprise. Therefore, the application or interpretation of these indicators should be accompanied (or replaced) by additional statistical indicators and methods. Because the human eye is a powerful pattern recognizer, graphical techniques and visualization methods for data characterization offer a simple yet powerful extension to the calculation of statistical moments [41].

6.6.3 Graphical Methods for Statistical Modeling: Frequency and Cumulative Frequency Distributions

Think of rearranging the measured samples of C_{ds} given in Table 6.3 in rank (ascending order). Next, assign *class intervals* (bins) of equal width over the range of the sample values and tally the number of samples in each bin. Using this method, either a *frequency histogram* or a *frequency polygon* (a line graph with points defined by the top right corners of the frequency histogram bins) can be created. The frequency histogram for the measured capacitor data in Table 6.3 is given in Figure 6.3. By visualizing the distribution in this way, it is clear that the

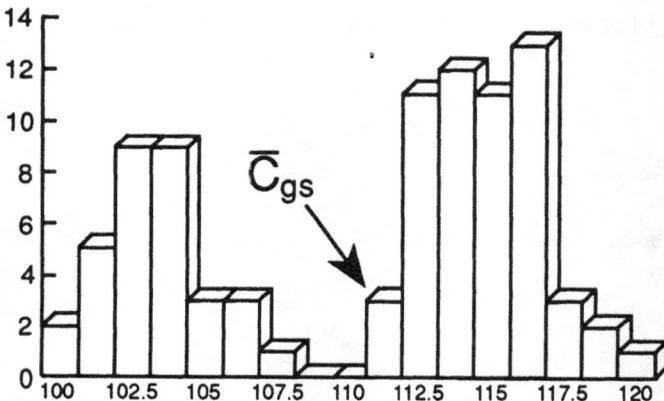

Figure 6.3 Frequency histogram for the FET parameter C_{ds}.

mean value of C_{ds} is a poor indicator of central tendency (in fact, it comes close to indicating that the most likely value of C_{ds} is that value of C_{ds} which can rarely occur). In the hallmark Bell System Technical Journal on statistical design [12], Logan remarked, "it would be meaningless to use some form of average for each of the transistor parameters as this could well result in a physically impossible combination of parameters." The warning here is "do not be tempted to use a statistical indicator until you are (reasonably) sure that it makes sense to do so in the given application." Perhaps the most obvious oversight in much of the existing work on device parameter statistical modeling is the reluctance of the investigator to assess the appropriateness of the statistical methods used in a particular situation.

As a final observation related to statistical moments, notice that the uniqueness of a distribution (even a univariate one) cannot be inferred from its mean value alone. This observation leads one to wonder how many moments are necessary to uniquely characterize a random parameter? In fact, all moments are not sufficient to fully describe a random variable [38]. Just the same, the moments have been used extensively to characterize device parameter databases. But even to use moments to approximately characterize database statistics has many problems:

1. There are too many of them.
2. We only have access to estimates of them.
3. It is difficult (if not impossible) to simulate distributions from moment information only.

To this end, we seek a compact and accurate representation of the multivariate frequency distribution, based on the sampled data, and which is not dependent upon calculated moments.

6.6.4 Truth Model

The truth model (TM) is a database model which provides a simple solution to the multidimensional statistical-modeling problem. It consists of a database of measured device-parameter samples, and can be thought of as a multivariate discrete density approximation to the overall continuous density function. This form of a "model" works well for statistical design because yield prediction and optimization using Monte Carlo techniques use discrete data points as trials.

To demonstrate the use of a statistically valid model, we present a design-centering example where a small-signal amplifier is centered: (1) using a moment-based device statistical model, and (2) with a measurement database statistical model.

6.6.5 Design Centering, Yield, and the Truth Model

The following example circuit is taken from [10]. The circuit and optimization specifications for this example are given in Figure 6.4. Starting from the single-

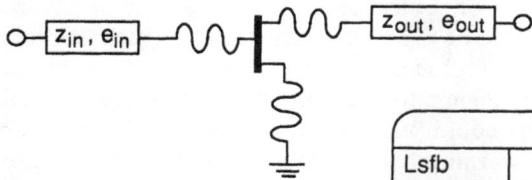

Optimization specs			
Lsfb	S_{11}	S_{22}	S_{21}
Nominal	< -10 dB	< -10 dB	= 15 dB
Yield	< -8 dB	< -8 dB	< 16, > 14 dB

Figure 6.4 Single FET 3.8- to 4.2-GHz amplifier used in the design-centering example.

point optimized component values as reported in [10], design centering using a center-of-gravity method was performed twice using the following assumptions on the FET model parameter statistics: (1) use the truth model, which in this case is a database of measured extracted FET circuit model parameters taken from FETs that were manufactured in the same manner as the circuit, and (2) use a low-order statistical model where distributions and pair-wise correlations are equivalent to that of the measured data (see Section 6.5.4). The designable parameters are z_{in}, z_{out}, e_{in}, e_{out}, ℓ_{in}, ℓ_{out}, ℓ_{sfb}, and they are modeled as independent uniform variables with $\pm 10\%$ tolerance limits. Tables 6.6 and 6.7 summarize the results of this centering exercise. Table 6.6 shows yield-estimate results (using a 5000-trial Monte Carlo analysis). Note that yield is improved by design centering when using either the low-order FET statistical model (31.4%–48.5%) or the truth model (31.4%–58.5%). But Table 6.7 shows that the results of the design centering procedure are affected by the statistical model used during centering. Remember

Table 6.6
Yield Estimates Before and After Design Centering

Design	Parameter Assumption for Yield Analysis	Yield Estimate Before Centering	Yield Estimate After Centering
1	Distributed-Correllated	44.6%	57.5%
1	Truth Model	31.4%	48.5%
2	Distributed-Correlated	44.6%	46.2%
2	Truth Model	31.4%	58.5%

Note: Design 1: Centered using Dist.-Corr. FET Model
Design 2: Centered using the Truth FET Model

Table 6.7
Design-Centering Results for the Single FET Amplifier

Parameter	Univariate Characterization	Multivariate Characterization	Delta %
z_{in}	35.52 Ω	38.15 Ω	7.87
e_{in}	83.78 deg	84.97 deg	1.42
ℓ_{in}	3.21 nH	3.23 nH	2.43
ℓ_{sfb}	0.55 nH	0.57 nH	4.56
ℓ_{out}	7.88 nH	8.48 nH	7.56
z_{out}	82.07 Ω	86.05 Ω	4.85
e_{out}	93.60 deg	99.80 deg	6.62

that the two models used in this example have statistically equivalent low-order univariate statistics and correlations. Yet, due to the multivariate higher order statistical differences between the two statistical models, both the yield and the design center values are affected by the model choice.

There is an important lesson to be learned here. The goal of design for manufacture is to design so there are no surprises during manufacture. The designer needs to be able to accurately calculate manufacturing yield, not just improve it. The designer needs to accurately determine the design center most suited to the manufacturing process. This example shows us two things:

- Design for manufacture requires accurate and properly validated statistical models of the manufacturing process.
- Matching the low-order statistics of the manufacturing and model statistics is generally not sufficient to assure a manufacturable design with no surprises.

Multivariate statistical validation of the manufacturing models, or using data base models, is a necessary step in design for manufacture.

6.6.6 Statistical-Interpolation Model

A FET is commonly characterized by a database containing a number of measurements, n, of actual manufactured FETs. For instance, 179 FET measurements were used in [33]. This database characterizes the statistics of the manufactured FET. A database model proposes to simply use the actual device parameter data when performing statistical analysis and design. This is practical and accurate when the number of measurements is "sufficient." However, a model which uses fewer measurements, the *statistical-interpolation model* (SIM), was recently proposed [33].

The technique used for the statistical-interpolation model is based on kernel-density estimation [40, 41]. In kernel-density estimation, data samples are used as the basis for defining the shape of a probability density function (PDF) which is used to model the PDF of the process from which the original data were generated. Model parameters are chosen so that the PDF of the model smooths or interpolates the data, while simultaneously matching the statistics of the data PDF.

The model is based on the following equation

$$\hat{S}_j = S_i + a\Delta S_j \, diag(K_i(k,h)) \tag{6.6}$$

where:

- $\hat{S}_j$ is the S-parameter vector generated from this model.
- S_i is a FET measurement vector chosen at random from the measured data.
- a is a constant model parameter.
- ΔS_j is a sample vector chosen at random from the kernel PDF.
- $K_i(k,h)$ is a scaling vector containing the distance from the chosen S_i to the kth nearest neighbor in each of the dimensions.

The kernel PDF for one example [33] is a 40-dimension, standard normal distribution with uncorrelated components and zero mean. The measured data from each FET is stored as a vector with 40 components (i.e., 8 parameters × 5 frequencies = 40 total). Lumping all the measurements over frequency into a single measurement vector is a strength, because the correlations and high-order statistical relations among the 40 parameters will be modeled when the entire vector is modeled. The spread around each data point, S_i, is determined by the model parameters "a" and "K_i." The sum of all the kernel PDFs forms the PDF used to generate the simulated data.

The choice of model parameters affects the smoothness of the simulated density function. For example, as seen in Figure 6.5 for a one-dimensional model, too small of an "a" value will cause too much granularity in the simulated PDF, whereas too large of an "a" value will cause the simulated PDF to be too smooth.

Model parameters are chosen so that the correlation matrices, and the marginal densities of the measured and simulated data match well. Analytical work shows that this model preserves the correlation matrix of the data only if the "a" variable is small.

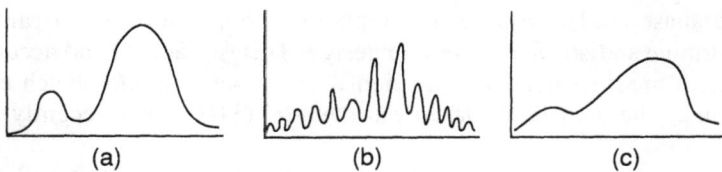

Figure 6.5 Density functions for a one-dimensional example showing (a) the original data PDF, (b) the simulated PDF when the "a" constant is too small, and (c) the simulated PDF when the "a" constant is too large.

Figures 6.6 and 6.7 illustrate the model results with some example two-dimensional scatter plots. These scatter plots were generated using the 179 measured FETs, and 10,000 simulated FET measurements, generated using the statistical-interpolation model. The measured data is shown on the left and the simulated data is shown on the right. Marginal densities for the simulated outcomes compare well to the marginal densities for the measured samples. All K-S numbers are greater than 0.98 with most of them being greater than 0.999. This shows that there is a 98% to 99.9% probability that the marginal densities match. The correlation matrices were compared using two methods: first, by computing the maximum difference between the elements of the matrices, and second, by the Euclidean norm of the difference matrix. The maximum difference was 0.06, and the Euclidean norm gave a value of 8.82e-4.

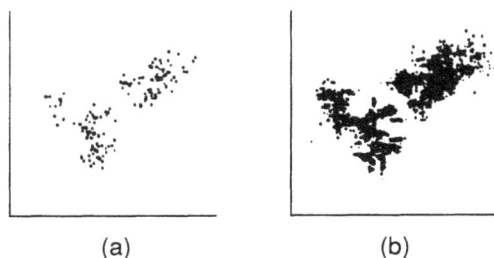

(a)   (b)

Figure 6.6 Re[S11] 6 GHz (X-axis) versus Im[S11] 6 GHz (Y-axis) for (a) measured and (b) simulated.

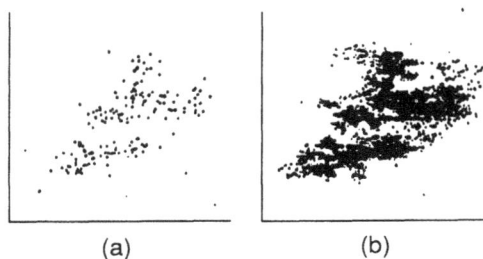

(a)   (b)

Figure 6.7 Im[S12] 1 GHz (X-axis) versus Re[S11] 26 GHz (Y-axis) for (a) measured and (b) simulated.

6.6.7 Summary of Statistical-Model Development

We have demonstrated that accurate design centering and yield prediction require accurate statistical models. In most cases, statistical equivalence (i.e., multivariate statistical equality with high confidence) between measured and simulated data is required to assure accurate statistical-design and yield-prediction results. Models based on low-order moments, such as means, variances, marginal densities, and pair-wise correlations do not generally provide statistical equivalence and are not recommended for accurate statistical design. The effects of poor statistical models on statistical-design and yield-prediction results for a simple microwave amplifier were examined. Additionally, we presented two recent device parameter statistical models which do provide statistical equivalence: the truth model, and the statistical-interpolation model. Next we incorporate the results of the previous sections into a recommended framework for general, accurate statistical modeling.

6.7 PROPOSED FRAMEWORK FOR STATISTICAL MODELING

A proposed framework for generalized statistical modeling is depicted in the flow diagram of Figure 6.8. We use the methodology from device modeling with statistically appropriate counterparts for characterization, modeling, extraction, and verification.

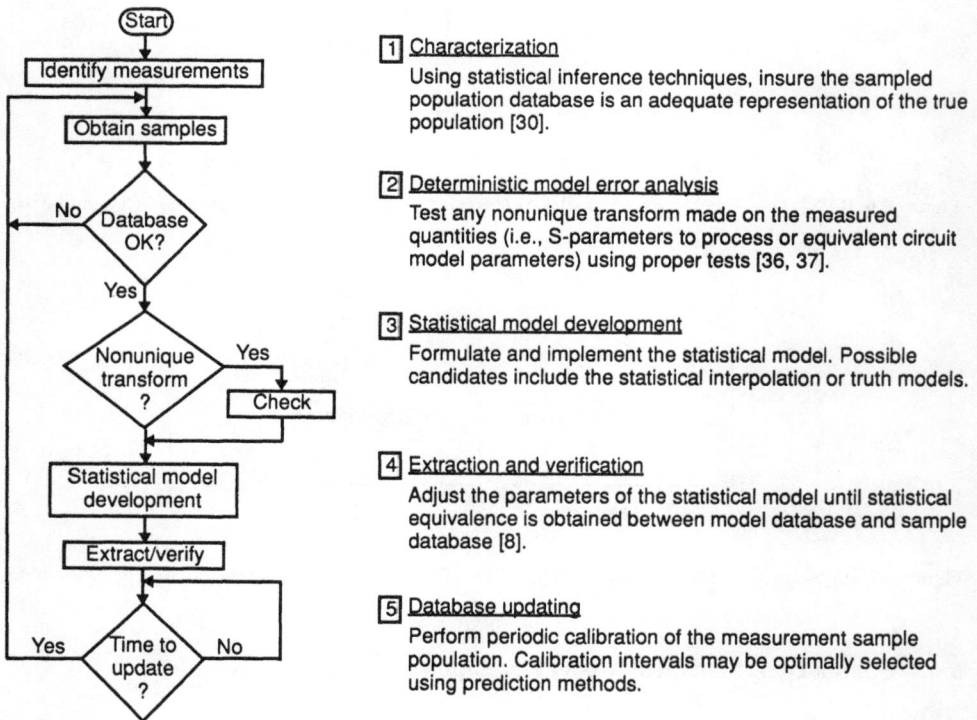

1 Characterization
Using statistical inference techniques, insure the sampled population database is an adequate representation of the true population [30].

2 Deterministic model error analysis
Test any nonunique transform made on the measured quantities (i.e., S-parameters to process or equivalent circuit model parameters) using proper tests [36, 37].

3 Statistical model development
Formulate and implement the statistical model. Possible candidates include the statistical interpolation or truth models.

4 Extraction and verification
Adjust the parameters of the statistical model until statistical equivalence is obtained between model database and sample database [8].

5 Database updating
Perform periodic calibration of the measurement sample population. Calibration intervals may be optimally selected using prediction methods.

Figure 6.8 Flow diagram of the framework for general, accurate statistical modeling.

6.7.1 Step 1: Characterization

Referring to the framework flow diagram, the statistical-modeling sequence begins with characterization, where the primary concern is the trade-off between database accuracy (size) and cost. We want to be reasonably sure that our sample represents the true population, subject to the cost of accomplishing this task. To properly choose the size of the database, we develop a scheme based on sampling theory which is linked to the sampling error in the yield estimate. Recall the Monte Carlo yield-estimate equation:

$$\hat{Y} = \frac{1}{N}\sum_{i=1}^{N} accept(P_i) \tag{6.7}$$

The mathematics of Monte Carlo tells us that the confidence we can have in the yield estimate, $\hat{Y}$, is a function of the number of trials, N, used in the Monte Carlo yield estimate. With this in mind we can formulate a sampling theory which answers the characterization problem of sizing the database. Table 6.8 shows the number of points required in a Monte Carlo yield calculation for a 50% yield (50% yield implies worst case values for N; i.e., the largest), for different confidence levels and confidence limits (see Section 3.3.1).

Table 6.8 is used as follows. If a ±5% error range is acceptable with a 95% confidence, then 384 sample points should be taken to characterize the process. A 95% confidence with a ±5% error means that the yield number calculated will be within ±5% of the actual yield 95% of the time that this calculation is made, assuming each time the calculation is made using a new sampling of the process.

Table 6.8
The Number of Monte Carlo Trials Needed for a Given Confidence in the Result, Assuming the Yield Is 50%

Error ±%	Confidence = 68.3%	Confidence = 95%	Confidence = 99%
1.0	2500	9604	16576
2.0	625	2401	4144
3.0	277	1067	1841
4.0	156	600	1036
5.0	100	384	663
6.0	69	266	460
7.0	51	196	338
8.0	39	150	259
9.0	30	118	204
10.0	25	96	165

Table 6.8 shows that many data points are needed to characterize the device parameter statistics if we desire accurate yield estimates. This sampling-theory-based method should help to make the trade-off between database size (i.e., the cost of characterization), and the resulting accuracy in the yield calculations. (The authors note that they have had reasonable practical success with 50 to 100 device parameter samples. But the main theme here is "more is better.")

6.7.2 Step 2: Deterministic-Model Error Analysis

The next step in device statistical modeling is to statistically quantify the influence of deterministic modeling errors resulting from any nonunique transformation on the original measured data. For example, the transformation from S-parameters to FET model parameters (these could be FET equivalent circuit or physical model parameters), and then back to S-parameters represents such a nonunique transformation. Figure 6.9 illustrates application of the statistical-equivalence test [36, 37] to determine the effects of FET modeling errors on database statistics. S-parameters are measured and then FET model parameters are extracted for each device. Next, each extracted FET is simulated to obtain simulated S-parameters. Finally, we test for statistical equivalence between the S-parameter data sets. If the test indicates statistical equivalence, then the effects of device modeling errors on the device statistics can be assumed negligible. (Obviously, the use of a device model could result in a substantial savings in the amount of data stored. Further, the use of physically based models could result in significant savings in the characterization effort.) The results of this step allow us to classify a deterministic model (i.e., device model) as "statistically valid." This type of study was performed

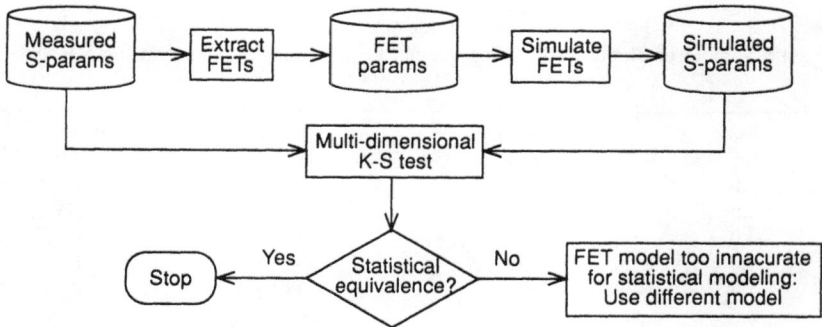

Figure 6.9 Flow diagram for testing how a nonunique transformation affects sample statistics.

in [6 and 28] except using low-order statistical tests, (i.e., means, variances and correlations) for verification.

6.7.3 Step 3: Statistical-Model Development

In step 3 the statistical model is chosen. The model should be able to accommodate any level of statistical complexity found in the sampled device parameter measurement database. Because of inherent accuracy and simplicity, the truth model (see Section 6.6.4) or the statistical interpolation model (see Section 6.6.6) for d-dimensional statistical data are presently our recommended statistical models. Either model can accommodate and accurately represent any sampled data set. Because the TM (or SIM) retains a high degree of statistical accuracy, small (typical) errors due to measurement or deterministic modeling will not manifest themselves into large ones when the model is used for simulation. Another benefit regarding either model is that they may be scaled, resulting in a reduction in effort required in the characterization step. Because it inherently interpolates, the potential for further savings in the characterization effort make the SIM very attractive. (However, the number of samples to use in the extraction of the SIM parameters is presently not clear.) Finally, note that since the TM is based only on the sampled population, there is no need to extract or verify it, again resulting in additional simplicity.

6.7.4 Step 4: Extraction and Verification

After characterization and modeling, our framework employs nonlinear optimization techniques in an effort to extract the SIM model parameters. In a way similar to the device modeling process, we are actually carrying out the verification step within the extraction step. The error-function formulation consists of the test for statistical equivalence (i.e., the nearest-neighbor-test statistic; see Section 6.5.3) applied to the sample measurements and to the generated outcomes from the SIM model.

6.7.5 Step 5: Database Updating

We incorporate periodic calibration of the model used in the framework to detect and correct any time-dependent shifts in the underlying population. In its simplest form, a moving window is applied so that old data is discarded when new data is available. Again, application of the statistical-equivalence test [36, 37] can be used to investigate the similarity between statistical models calibrated at different times.

Since the TM (or SIM) model is scalable, its parameters can be predicted to give insight as to how the population is changing with time. This information could then be used to help schedule recalibration of the database.

6.8 CONCLUSION

In this chapter we have examined the device-parameter statistical-modeling problem, and various proposals for its solution. Low-order, univariate statistical moments were shown to be of little use for discriminating between like and unlike multivariate data sets. Motivated by this, a comprehensive framework for generalized, accurate statistical modeling was presented. Composed within the familiar methodology used in device modeling, statistically appropriate counterparts for characterization, modeling, extraction, and verification were examined. The generality of the framework is insured by adopting statistical models and verification techniques capable of dealing with arbitrarily distributed joint density characteristics. Due to this generality, the framework can easily accommodate any type of random parameters, from systems to physical-process parameters. Finally, the statistically powerful verification techniques used herein provide a basis by which any device statistical-model formulation can be compared.

6.9 IMPORTANT IDEAS FROM CHAPTER 6

Section 6.1

- If the model of the underlying parameter statistics is inaccurate, then it is likely that design results will be inferior.

Section 6.2

- A key theme in much of the literature is the need both to describe and to generate outcomes for a multivariate joint density. However, not until recently have explicit multivariate methods been used in this endeavor.
- Univariate methods, such as comparing means, correlations and marginal densities, are not sufficient tests to validate a multivariate statistical model.

Section 6.3

- Statistical model development, both deterministic and statistical, involves four steps:
  1. Characterization;
  2. Model development and extraction;
  3. Verification;
  4. Simulation.

Section 6.4

* Characterization involves two steps:
  1. Determining what to measure;
  2. Taking measurements.
* The key to statistical modeling is to discover a model whose outcomes create a simulated database which is statistically equivalent to the measured device parameter database.

Section 6.5

* The verification step in statistical modeling measures the accuracy of the statistical model used.
* Statistical equivalence between two databases occurs when both databases represent samples from the same PDF with a high degree of confidence.
* Two tests for multivariate statistical equivalence are introduced:
  1. The generalized Kolmogorov-Smirnov test;
  2. The nearest neighbor test.

Section 6.6

* Statistical-model development usually involves determining statistical relations within the measurement database, and developing analytic expressions which can reproduce the relations in simulated data.
* Using the "average" device as a statistical model can lead to surprises when the unit is manufactured.
* To use any statistical moments blindly, without regard for the underlying PDF of the sample is an invitation for surprise.
* Two possible candidates statistically valid models for statistical design and analysis are:
  1. The truth model;
  2. The statistical-interpolation model.

Section 6.7

* A generalized framework for statistical modeling is presented in this section.
* A sampling theory based on Monte Carlo trials is developed and should help with the trade-off between database size and the resulting accuracy (i.e., confidence) of the resulting statistical calculation.

REFERENCES

[1] J. Purviance and M.D. Meehan, "CAD for Statistical Analysis and Design of Microwave Circuits," Int. Jour. of Micro. and Mill.-Wave C.-A. Engr., Vol.1, No.1, 59-76, Jan. 1991.
[2] J. W. Bandler, and S.H. Chen, "Circuit Optimization: The State of the Art", IEEE Transactions on Microwave Theory and Techniques, Vol 36, No. 2, Feb. 1988, pp. 424-42.

[3] R. Spence and R. Soin, *Tolerance Design of Electronic Circuits*, Addison-Wesley Publishing Company, 1988.

[4] R.K. Brayton, G.D. Hachtel, and A.L. Sangiovanni-Vincentelli, "A Survey of Optimization Techniques for Integrated Circuit Design," Proc. IEEE, Vol 69, No. 10, pp. 1334-1363, 1981.

[5] R. Brayton, R. Spence, *Sensitivity and Optimization*, Elsevier, New York, 1980.

[6] R.Anholt, R.Worley, and R.Neidhard, "Statistical Analysis of GaAs MESFET S-Parameter Equivalent-circuit Models," International Journal of Microwave and Millimeter-Wave Computer-Aided Design, Vol. 1, No.3, July 1991, pp. 263-270.

[7] M. Meehan, T. Wandinger, and D. Fisher, "Accurate Design Centering and Yield Prediction Using the "Truth Model," Proceedings of the 1991 IEEE MTT-S Int. Microwave Symposium, June 1991, pp. 1201-1204.

[8] M. Meehan and L. Campbell, "Statistical Techniques for Objective Characterization of Microwave Device Statistical Data," Proceedings of the 1991 MTT-S Int. Microwave Symposium, June 1991, pp 1209-1212.

[9] J. Purviance, M. Meehan, and D. Collins, "Properties of FET Statistical Databases", Proceedings of the 1990 IEEE MTT-S International Microwave Symposium, May 1990.

[10] J. Purviance, D. Criss, and D. Monteith, "FET Model Statistics and Their Effects on Design Centering and Yield Prediction for Microwave Amplifiers," Proceedings of the 1989 IEEE MTT-S International Microwave Symposium, June 1989.

[11] M.D. Meehan and D.M. Collins, "Investigations of the GaAs FET Model to Assess its Applicability to Design Centering and Yield Estimation," EEsof Development Report, Dec. 1987.

[12] J. Logan, "Characterization and Modeling for Statistical Design," The Bell System Technical Journal, Vol. 50, No. 4, 1971, pp. 1105-1147.

[13] P.J. Rankin, "Statistical Modeling for Integrated Circuits," IEE Proceedings, Vol. 129, Pt.G, No. 4, aug 1982, pp 186-191.

[14] S. Inohira, T. Shinmi, M. Nagata, T. Toyabe, and K. Iida, "A Statistical Model Including Parameter Matching for Analog Integrated Circuits Simulation," IEEE Transactions on CAD, Vol. CAD-4, No.4, Oct. 1985, pp.621-628.

[15] A.J. Strojwas, "Design for Manufacturability and Yield," 26th ACM/IEEE Design Automation Conference, Paper 29.1.

[16] C. Ho, S. Hansen, and P. Fahey, "SUPREM III—A Program for Integrated Circuit Modeling and Simulation," Tech. report, Stanford University, 1984.

[17] Z. Yu and R. Dutton, "Sedan III—A Generalized Electronic Material Device Analysis Program," Tech Report, Stanford University, 1985.

[18] W. Maly and A. Strojwas, "Statistical Simulation of the IC Manufacturing Process," IEEE Transactions on Computer-Aided Design, Vol. CAD-1, July 1982, pp. 120-131.

[19] S.R. Nassif, A.J. Strojwas, and S.W. Director, 'FABRICS II; A Statistically Based IC Fabrication Process Simulator," IEEE Transactions on Computer-Aided Design, Vol. CAD-3, Jan. 1984, pp.40-46.

[20] C. Spanos, S.W. Director, "Parameter Extraction for Statistical IC Process Characterization," IEEE Transactions on Computer-Aided Design, Vol. CAD-5, No. 1, Jan. 1986, pp. 66-78.

[21] C. K. Chow, "Statistical Circuit Simulation of a Wideband Amplifier: A Case Study in Design for Manufacturability," Hewlett-Packard Journal, Oct. 1990, pp. 78-81.

[22] J. P. Spoto, W. T. Coston, and C. P. Hernandez, "Statistical Integrated Circuit Design and Characterization," IEEE Transactions on Computer-Aided Design, Vol. CAD-5, No. 1, Jan. 1986, pp. 90-103.

[23] S. Liu, K. Singhal, "A Statistical Model for MOSFETS," IEEE International Conference on Computer-Aided Design (ICCAD), Nov. 1985, pp. 78-80.

[24] N. Herr and J.J. Barnes, "Statistical Modeling for Circuit Simulation of CMOS VLSI," IEEE International Conference on Computer-Aided Design (ICCAD), Nov. 1985, pp. 81-83.

[25] J. Purviance, M. Petzold, and C. Potratz, "A Linear Statistical FET Model Using Principal Component Analysis," IEEE Transactions on Microwave Theory and Techniques, Vol. 37, No. 9, Sept. 1989, pp. 1389-1394.

[26] *Microwave Engineering Europe,* July/August 1990, pp. 20.

[27] R. Anholt, J. King, R. Worley, and J. Gillespie, "Relationship between Process and Materials Variations and Variations in S- and Equivalent-Circuit Parameters," International Journal of Microwave and Millimeter-wave Computer-Aided Design, Vol. 1, No. 3, July. 1991, pp. 271-281.

[28] J.W. Bandler, R.M. Biernacki, S.H. Chen, J. Song, S. Ye, and Q.J. Zhang, "Statistical Modeling of GaAs MESFETs," IEEE MTT-S International Microwave Symposium, Boston, MA., June 1991, pp. 87-90.

[29] R.J. Gilmore, M. Eron, and T. Zhang, "Yield Optimization of a MMIC Distributed Amplifier using Physically-Based Device Models," IEEE MTT-S International Microwave Symposium, Boston, MA., June 1991, pp. 1205-1208.

[30] W.J. Conover, Practical Nonparametric Statistics, John Wiley & Sons, Inc., New York, 1980.

[31] J.W. Bandler, R.M. Biernacki, S.H. Chen, J.F. Loman, M.L. Renault, and Q.J. Zhang, "Combined Discrete/Normal Statistical Modeling of Microwave Devices," Proceedings of the 19th European Microwave Conference, Sept. 1989, pp. 205-210.

[32] M.K. Kalos and P.A. Whitlock, Monte Carlo Methods, Volume 1: Basics, Wiley and Sons, 1986.

[33] L. Campbell, J. Purviance, C. Potratz, "Statistical Interpolation of FET Database Measurements," IEEE MTT-S International Microwave Symposium, Boston, MA., June 1991, pp. 201-204.

[34] D.L. Allen, J. Beal, M. King, "Small-signal RF Yield Analysis of MMIC Circuits Based on Physical Device Parameters," IEEE MTT-S International Microwave Symposium, Albuquerque, NM, June 1992, pp. 1473-1476.

[35] J.H. Friedman and L.C. Rafsky, "Multivariate Generalizations of the Wald-Wolfowitz and Smirnov Two-Sample Tests," The Annals of Statistics, Vol. 16, 1979, pp. 772-783.

[36] M.F. Schilling, "Multivariate Two-Sample test Based on Nearest Neighbors," Journal of the American Statistical Association, Vol. 81, No. 395, pp. 799-806, September 1986.

[37] N. Henze, "A Multivariate Two-Sample Test Based on the Number of Nearest Neighbor Type Coincidences," The Annals of Statistics, Vol. 7, 1988, pp. 697-717.

[38] L. Devroye, Non-Uniform Random Variate Generation, Springer-Verlag, 1986.

[39] Touchstone[a], Libra[a] and Omnisys[a] User's Manuals, EEsof Inc., Westlake Village, CA., 1991.

[40] L. Breiman, W. Meisel, and E. Purcell, "Variable Kernel Estimates of Multivariate Densities," Technometrics, Vol. 19, No. 2, May 1977, pp. 135-144.

[41] D. W. Scott, *Multivariate Density Estimation—Theory, Practice, and Visualization,* John Wiley & Sons, Inc., New York, 1992.

Chapter 7

Examples and Case Studies

"In order to determine and improve the production potential of a design, it must be analyzed and optimized over the entire range of variables that will be encountered during manufacturing."

"The fundamental truth in analysis and parameter modeling is simply that you usually get what you pay for—nothing more."

7.1 INTRODUCTION

This chapter presents examples of circuit and system design using methods and ideas presented in this book. This chapter serves two purposes. First, it provides the reader with a general methodology for practicing robust design in several unique design situations. Second, case studies from industry provide a contemporary sample of practical work in the area of statistical design and modeling.

7.2 EXAMPLE—A COMPREHENSIVE DESIGN USING A LOWPASS FILTER

7.2.1 Comments

In this example a two-section lumped-element RF lowpass-filter-design problem illustrates the brinkmanship tendency of single-point optimization. In addition, a methodology for statistical analysis and design is provided using a commercially available simulation and optimization package.

7.2.2 An Extended-Design Methodology

The application of statistical-design techniques can be viewed as a straightforward extension of most design methods currently in use. Figure 7.1 depicts this situation where the design solution due to single-point optimization is further refined to include realistic parameter variations. Setting the parameter variation is the subject of the first step in statistical design: modeling the unit parameter joint PDF. Statistical modeling was covered at length in Chapter 6, and can be as simple as setting parameter values to low, medium, and high, as is practiced in design of experiments, or as involved as the methods described in the case studies reviewed later in this chapter. The fundamental truth in analysis and parameter modeling is simply that "you usually get what you pay for—nothing more." For example, classical static mechanics can be used to analyze a bridge in a straightforward manner. However, without dynamic analysis, nobody knows what oscillatory tendencies the structure might exhibit during a severe storm. A simple analysis procedure and model will usually not give exact and detailed results.

Because it is desirable to know how the unit response will behave in the presence of random design parameters, the next step in statistical design is *statistical*

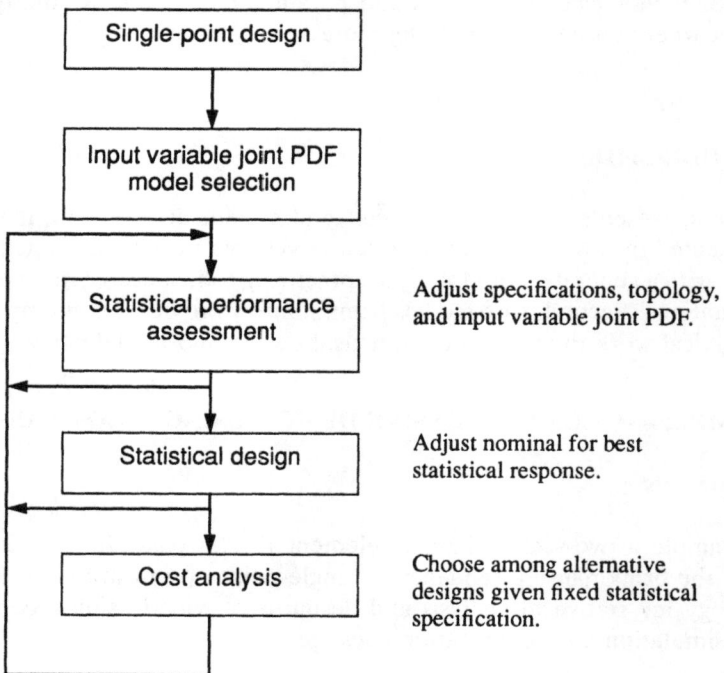

Figure 7.1 Flow diagram of an extended methodology for unit design that includes statistical design.

analysis. A response specification which may be used for a single-point design is usually relaxed when used to evaluate the unit statistical response. For example, specifications requiring equality are not appropriate for statistical design as they will generally imply unit failure. Also, like the bridge dynamics, the designer just does not know the capacity of the design to withstand manufacturing conditions until it is interrogated as such. In some respects, the single-point specification can be viewed as a "guideline" during initial design steps, and it is then superseded by a *statistical specification* once unit statistical performance has been characterized.

Measuring the unit statistical performance is usually the most difficult aspect of statistical design. It can often involve the iterative trading between random performance and statistical specification, and thus yield. It almost always results in relaxation of the single-point specification, sometimes the input parameter PDF (tighter tolerances), and can even prove that the underlying design topology is inadequate.

Once it appears reasonable that the selected unit topology and input PDF descriptions provide satisfactory statistical performance (when viewed against the statistical specification), the process of adjusting the nominal values of so-called "designable" parameters can begin. This is the "design" part of statistical analysis and design. This design stage makes use of the methods described in Chapter 5, and usually results in a refinement of the initial unit statistical specification. If the yield optimizer is successful, then the statistical-performance specification can be revised in a favorable direction, provided the cost due to any associated decrease in yield is not too large. Figure 7.2 depicts the statistical analysis and design process as a three-way balancing act between unit yield, cost, and performance. The majority of this book has been devoted to the tools required to determine the proper balance point or "best mix."

7.2.3 The Statistical-Design Methodology in Practice

This example was reported in [1]. It consists of a two-section RF lowpass filter with the single-point specifications given in List 7.1.

List 7.1 The Initial Single-Point Performance Specification

Filter bandwidth BW = 50 MHz
Pass-band ripple < 1 dB
Input return loss < −14 dB
Stop-band attenuation = −24 dB one octave above cut-off

Fabrication assumptions include the use of glass-epoxy board and standard-value discrete components. The schematic diagram of the filter is shown in Figure 7.3, which is a screen dump from a commercial circuit and system simulator [2]. The simulator provides an intuitive framework that streamlines the design engineer's design task. The framework uses a laboratory- and production-environment meta-

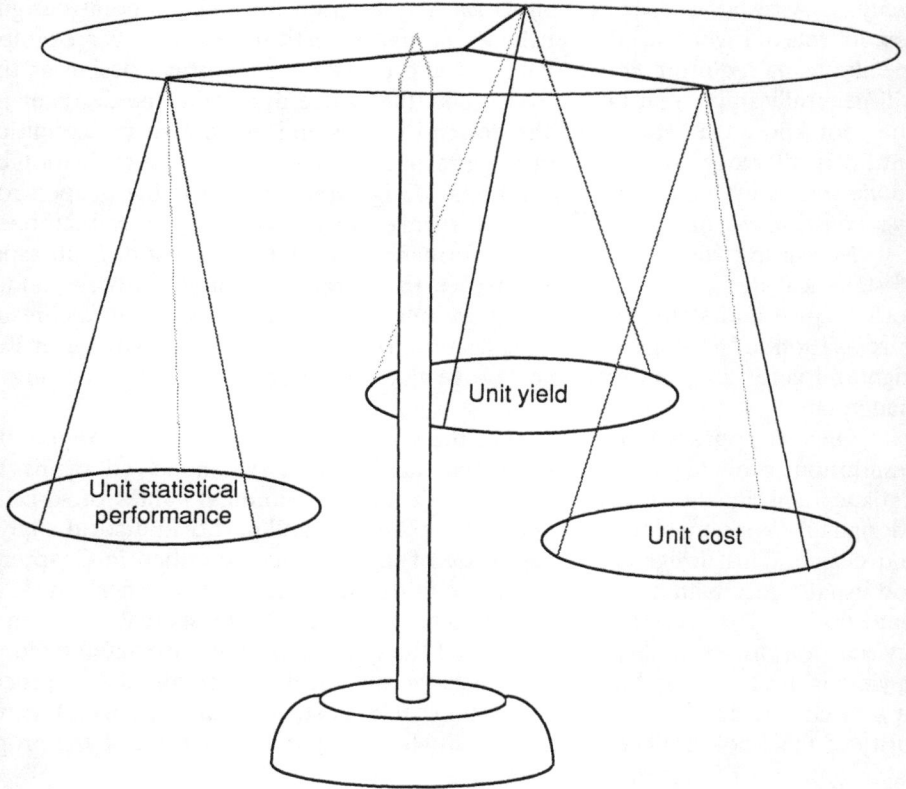

Figure 7.2 Statistical analysis and design—a three-way balancing act between yield, performance and cost.

phor where a device under test (DUT) or a device under production (DUP) can be directly manipulated with a computer-pointing device (mouse). The *test bench* depicted in the lower half of Figure 7.3 shows other objects such as meters for performing unit measurements, and *stimulus sources* for exciting the DUT. Additionally, plotting and graphing tools are available to help document the design's performance. This type of graphically-oriented design environment is becoming popular in the industry.

Statistical design for the lowpass filter begins after single-point gradient optimization terminates due (in this case) to sufficiently small (effectively zero) gradients. Because the optimizer was not able to satisfy the input return loss specification at the band edge (−11.5 dB at 50 MHz.), we already have an indication that the current topology may be inadequate for our purposes. Nonetheless, we use the optimal single-point parameter values given in List 7.2.

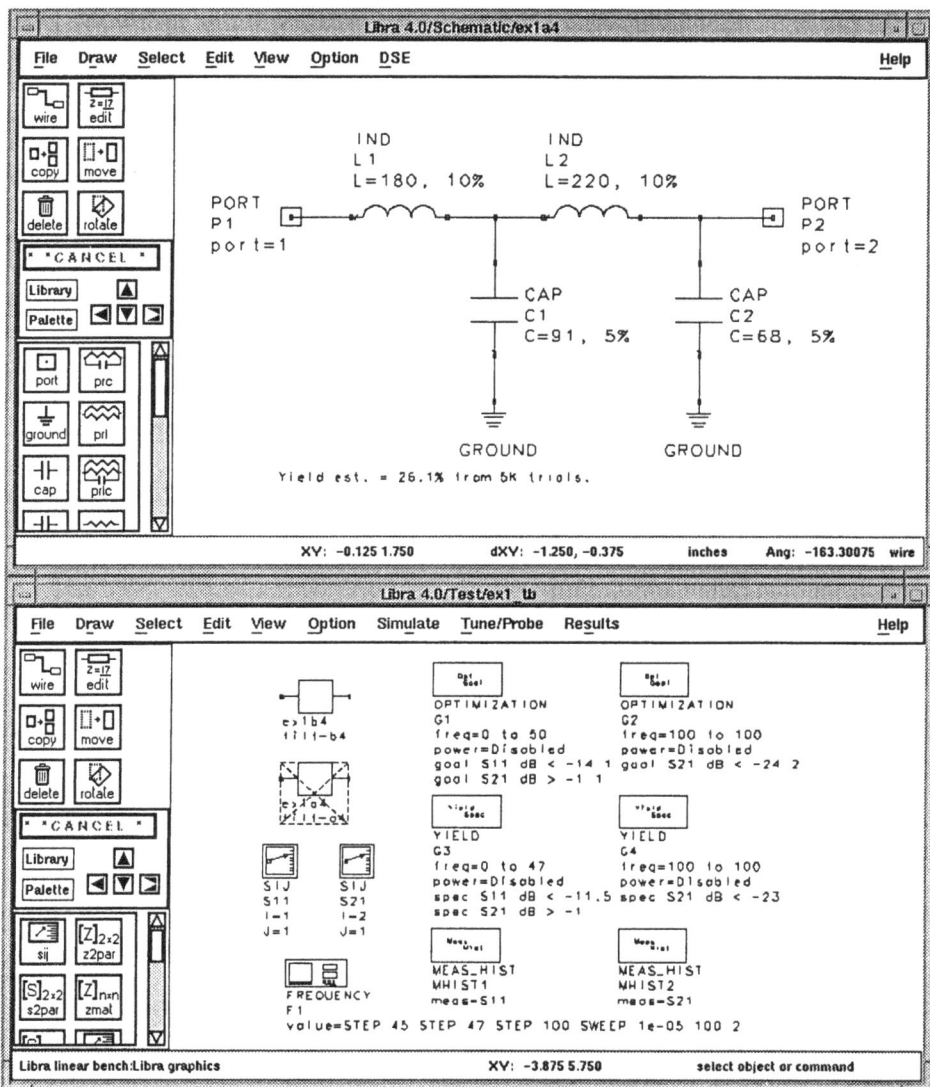

Figure 7.3 Screen dump of an advanced circuit- and system-design environment.

List 7.2 Single-Point Optimized Parameter Values

$L1 = 168.7$ nH
$C1 = 94.6$ pF
$L2 = 236.3$ nH
$C2 = 67.6$ pF

The simulation includes the frequencies 0 to 100 MHz in steps of 2 MHz, along with the band-edge frequency points of 45 and 47 MHz. Because simulation time increases with the number of discrete swept parameter points (e.g., frequency and power), care must be exercised to select a small set of frequency points which adequately capture the simulation response, particularly in regions of interest. Also note that since the response shape can change during optimization, so it may be necessary to periodically reallocate simulation points.

Now that the single-point optimal filter parameters have been determined, the statistical performance using the closest standard-value discrete parts may be obtained. The statistical model used here is independent uniformly distributed parameters. Figure 7.4 (a, c) show the two-sigma performance envelope using the parameter values given in List 7.3.

List 7.3 Standard Nominal Values Closest to the Single-Point Optimized Values From List 7.2

$L1 = 180$ nH $\pm 10\%$
$C1 = 91$ pF $\pm 5\%$
$L2 = 220$ nH $\pm 10\%$
$C2 = 68$ pF $\pm 5\%$
Independent and uniformly distributed

The two-sigma envelope plots are derived from the statistical response, and therefore bound approximately 95.4% of all performance outcomes. Thus if both envelopes were just within specification (or one within and the other just within), the yield would be 95.4% (subject to confidence calculations). Notice in Figure 7.4(c) that the single-point stop-band attenuation specification is approximately satisfied by the two-sigma performance.

A cautionary note on the performance-envelope plots is in order. Although several response plots may be presented for a design, it is not possible to determine the relation among the plots by observing individual plots. It is tempting to think that the parameter values that produce a poor response in one plot are the same parameters that produce the poor responses in the other plots. This is not generally true. One set of parameters may cause a satisfactory S11 but a poor S22, while another set of parameters can do just the opposite. Because there is no straightforward method for showing the relationship between multiple measurements, the overall interpretation of several statistical response plots must be handled with care. The relationship *between* measurements in either envelope or trace-type plots is not intuitive. For this purpose the overall yield estimate is best, and the envelope and trace plots can be used to examine trends before, during and after statistical design—they provide excellent feedback with regard to the effectiveness of any statistical-design method.

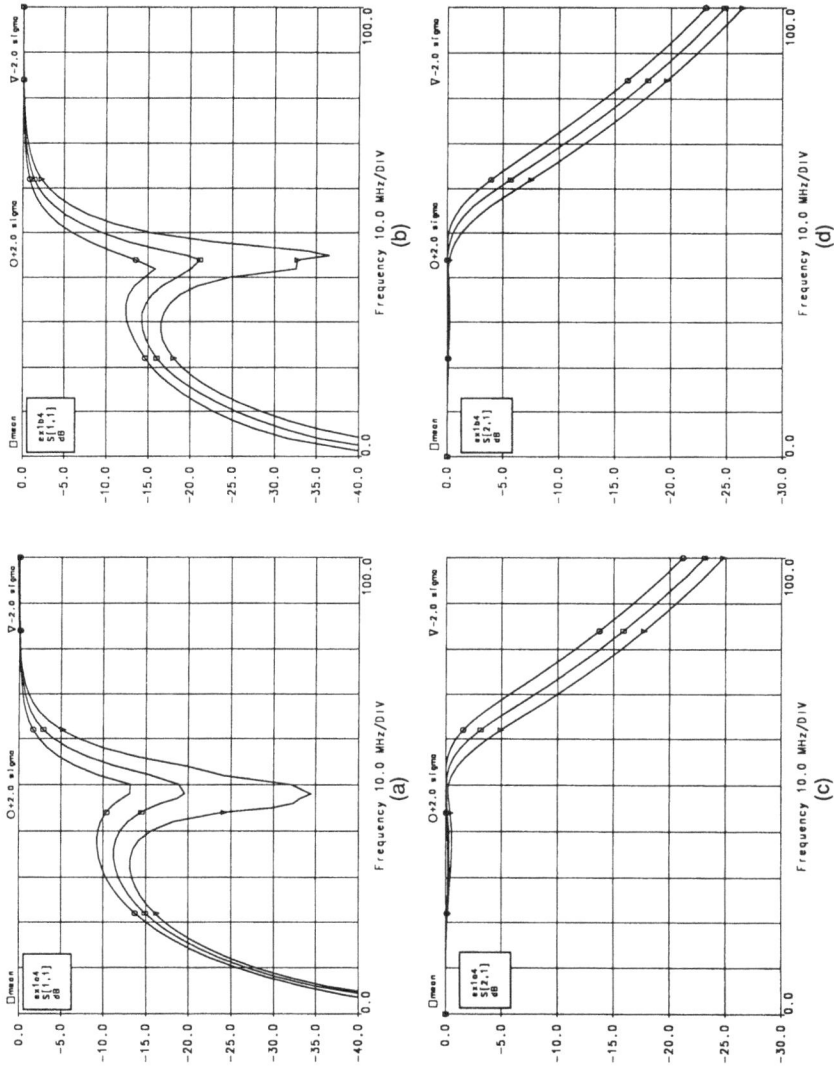

Figure 7.4 Two-sigma performance envelopes using: (a, c) the parameter values L1 = 180 nH ±10%, C1 = 91 pF ±5%, L2 = 220 nH ±10%, C2 = 68 pF ±5%, independent and uniformly distributed; and (b, d) the parameter values L1 = 180 nH ±10%, C1 = 91 pF ±5%, L2 = 270 nH ±10%, C2 = 62 pF ±5%.

Turning now to the yield estimate using the single-point performance specification from List 7.1, and the standard parameter values and tolerances found in List 7.3, we found the yield was negligible. Therefore, it is necessary to revise the specification as given in List 7.4.

List 7.4 The Statistical Performance Specification Based on the Single-Point Specification in List 7.1

Filter bandwidth BW = 47 MHz
Pass-band ripple < 1 dB
Input return loss < -11.5 dB
Stop-band attenuation = -23 dB one octave above cut-off

The yield estimate for the statistical specification is 26.1% based on 5,000 Monte Carlo trials. The yield estimate from the revised specification, along with the two-sigma envelope plots provide a better understanding of what can be expected from our production-bound filter.

With the revised specification in hand, the statistical-design process begins with an initial set of yield factor histograms (YFHs). Figure 7.5 shows the YFHs for the four parameters of the filter C_1, C_2, L_1, and L_2. Notice the strong yield gradients exhibited in C_1 and L_2 (Figures 7.5(a) and (d), respectively). We could use the YFHs directly to influence a new choice of nominal values which would hopefully increase the yield. But instead, we assume the nominal parameter values are continuous (rather than discrete standard values), and use one of the methods for yield optimization from Chapter 5. In this way we can obtain an upper bound on the yield to compare to the yield of the filter using standard nominal parameter values. Additionally, the optimal yield parameter values will aid in the selection of the off-the-shelf standard values. After design centering with a modified center-of-gravity approach, the yield is 89.9% (5,000 trials). The optimal yield parameter values are given in List 7.5.

List 7.5 Optimal Yield Nominal Parameter Values

L1 = 195.8 nH ±10%
C1 = 89.7 pF ±5%
L2 = 262.8 nH ±10%
C2 = 59.2 pF ±5%

The closest standard nominal values are given in List 7.6. These values produce a yield of 76.2% (5,000 trials).

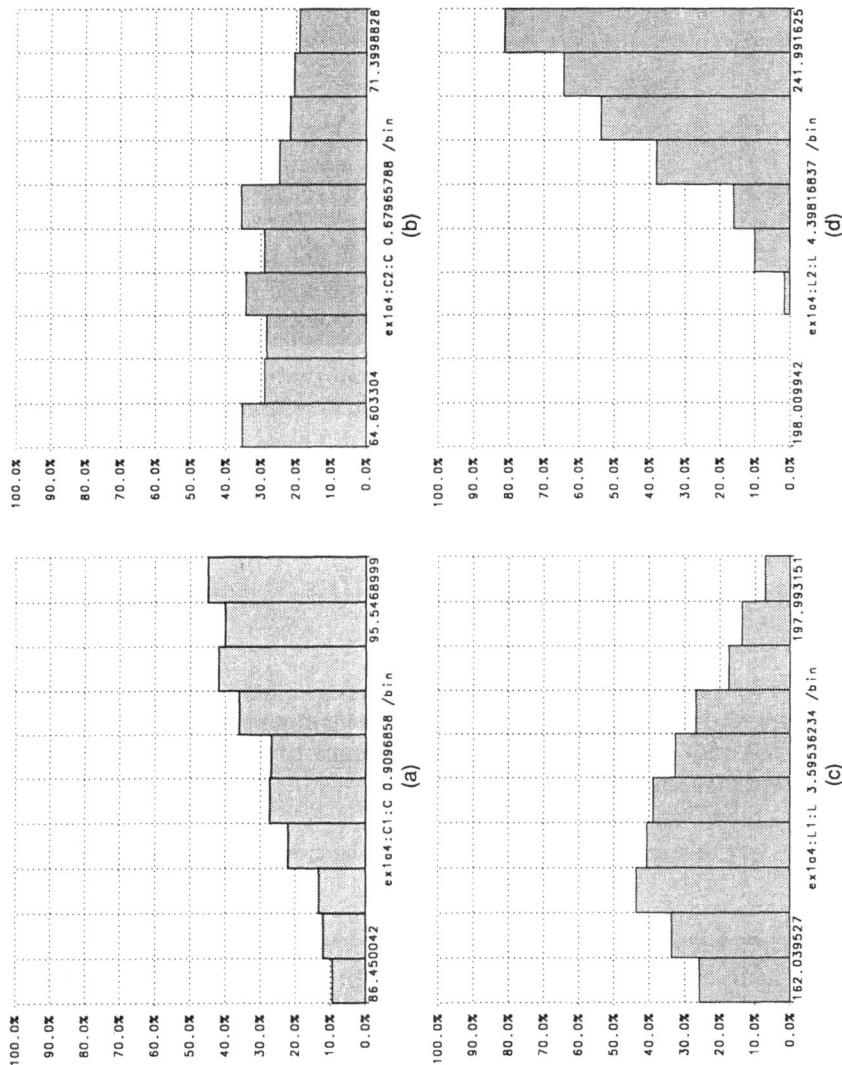

Figure 7.5 Yield factor histograms for the parameters: L1 = 180 nH ±10%, C1 = 91 pF ±5%, L2 = 220 nH ±10%, C2 = 68 pF ±5%, independent and uniformly distributed; and for the statistical specification: filter bandwidth BW = 47 MHz, pass-band ripple < 1 dB, input return loss < −11.5 dB, stop-band attenuation = −23 dB one octave above cut-off.

List 7.6 Standard Nominal Values Closest to the Yield-Optimized Values From List 7.5

L1 = 180 nH ±10%
C1 = 91 pF ±5%
L2 = 270 nH ±10%
C2 = 62 pF ±5%

The YFHs for this design appear in Figure 7.6, and the two-sigma response envelopes are shown in Figure 7.4(b, d). Notice the translation and compression of the S21 response at the stop-band edge (100 MHz.). Even though the single-point stop-band attenuation specification was relaxed, the statistical-design algorithm still improved this response characteristic. Additionally, the pass-band ripple envelope is further compressed. Finally, notice that the input return loss envelope is translated away from the −11.5 dB specification.

Lastly we examine the YFHs for the single-point optimized design and the yield-optimized design, where in both cases 10% inductors and 5% capacitors are assumed. Figure 7.7(a through d) shows the single-point YFHs, indicating that even when single-point gradients are effectively zero, there can be strong yield gradients—confirming the brinkmanship tendency of the single-point optimizer. In contrast, the yield-optimized YFHs (Figure 7.7e through h) are either flat or symmetric about the nominal.

7.2.4 Summation

Using a relatively simple circuit, a statistical circuit-design methodology was demonstrated. In order to determine and improve the production potential of a design, it must be analyzed and optimized over the entire range of variables that will be encountered during manufacture.

7.3 EXAMPLE—A 2- TO 6-GHz GaAs MMIC FEEDBACK AMPLIFIER

7.3.1 Comments

This example again shows the brinksmanship nature of single-point optimized designs, in this case using a minimax single-point optimizer. In this example only three of the eight statistical variables were changed in the optimization, yet there was still good yield improvement. The statistical models used here are very simple and are not validated.

7.3.2 Preliminary Information

In this example we consider a two-stage 2- to 6-GHz MMIC feedback amplifier which first appeared in [3] and then was used in [4]. The equivalent circuit model

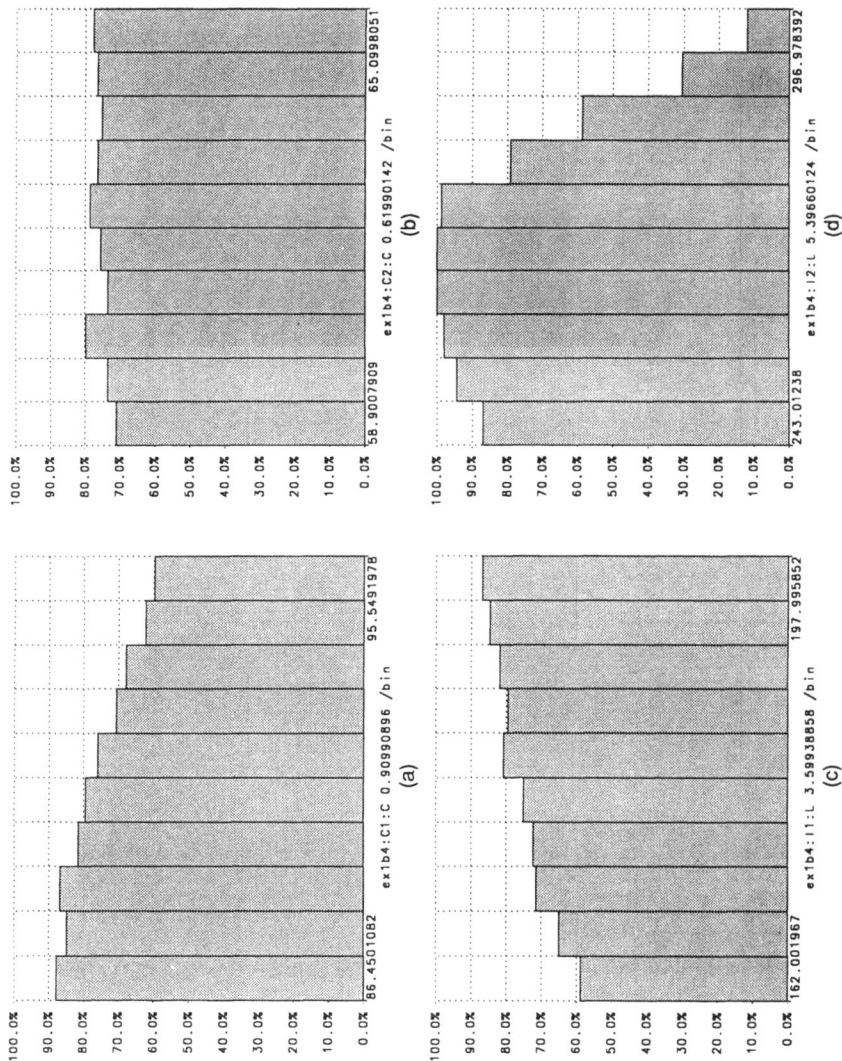

Figure 7.6 Yield factor histograms for the parameters: L1 = 180 nH ±10%, C1 = 91 pF ±5%, L2 = 270 nH ±10%, C2 = 62 pF ±5%; and for the statistical specification: filter bandwidth BW = 47 MHz, pass-band ripple < 1 dB, input return loss < −11.5 dB, stop-band attenuation = −23 dB one octave above cut-off.

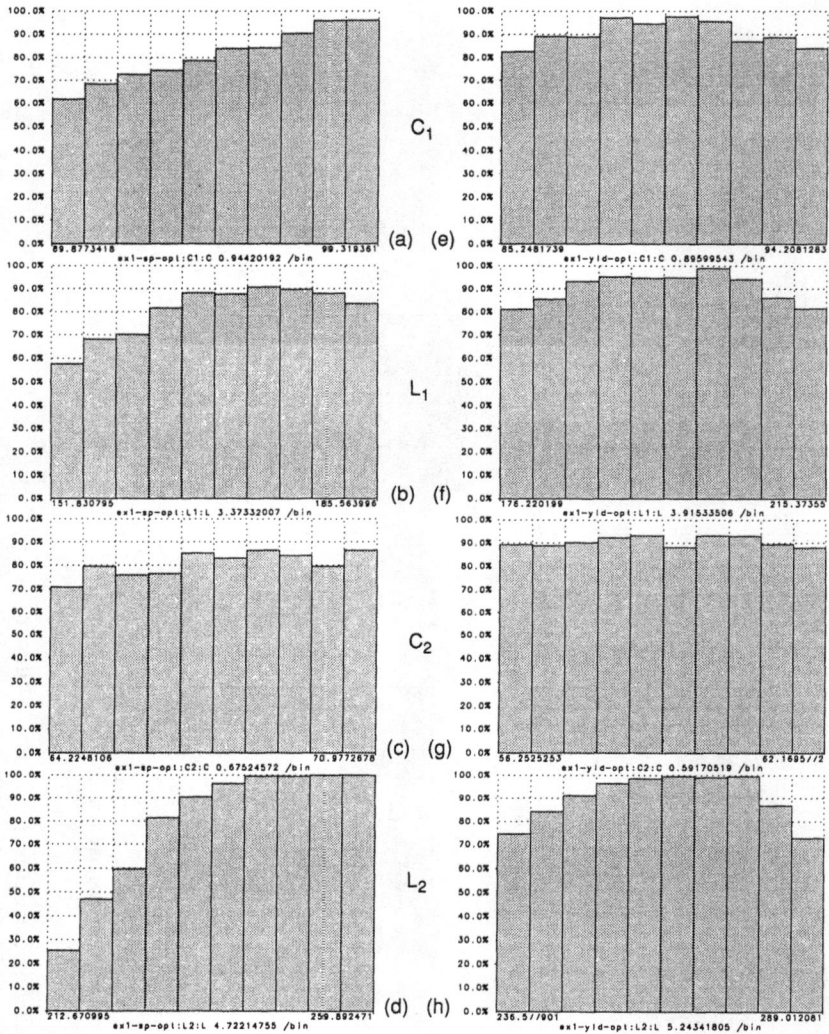

Figure 7.7 Yield factor histograms for (a through d) the single-point optimized design, and (e through h) the yield-optimized design.

for the FET and the circuit are shown in Figure 7.8. The specifications are a small-signal gain of 8 dB ± 1dB, VSWR at the input port (VSWR1) of less than 2, and VSWR at the output port (VSWR2) less than 2.2. A total of 9 sampling-frequency points equally spaced with the step of 0.5 GHz are used.

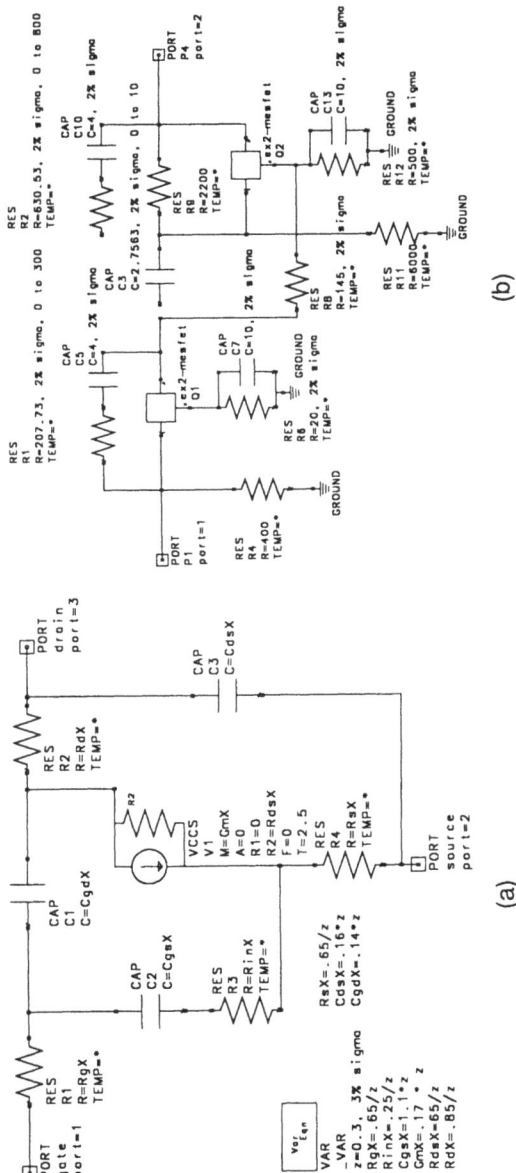

Figure 7.8 (a) Normalized GaAs MESFET model, Z is the gate width in millimeters, $G_m = 0.17Z$, and $r = 2.5$ ps. All resistors are in ohms. All capacitors are in picofarads. (b) A two-stage amplifier.

7.3.3 Statistical-Parameter Model

The statistical model used for the FET and circuit parameters is a simple one. The mean value of the capacitors is fixed to keep the size of the circuit reasonable. The mean value of the gate width is fixed because of the assumed FET process, but a 3% standard deviation is allowed. Since the RF responses are not very sensitive to changes in the bias resistors, no tolerances are assigned to the resistors. Two feedback resistors and interstage coupling capacitor are chosen as design variables. A standard deviation of 2% is assumed for the design variables. Independent Gaussian distributions are used for the parameter PDFs. The nominal values and standard deviations of the parameters are listed in Table 7.1.

Table 7.1
Parameter Values and Tolerances for the MIMIC Amplifier

Element Parameter	Nominal Value	Standard Deviation
Z (μm)	300	3%
R_4 (Ω)	400	0%
C_5 (pF)	4	2%
R_6 (Ω)	20	2%
C_7 (pF)	10	2%
R_8 (Ω)	145	2%
R_9 (Ω)	2200	0%
C_{10} (pF)	4	2%
R_{11} (Ω)	6000	0%
R_{12} (Ω)	500	2%
C_{13} (Ω)	10	2%

7.3.4 Statistical Optimization and Analysis

As a first step in the statistical optimization of this amplifier, a minimax single-point optimized solution is found as a starting point for the statistical optimization. The minimax solution is shown in Table 7.2. The yield estimate for the minimax design based on 5000 Monte Carlo trials is 29.3%.

Table 7.2
Yield-Optimization Results of the MMIC Amplifier Using a Center-of-Gravity Method

Parameter	Initial Values	Values After Statistical Optimization
R_1 (Ω)	201	201.4
R_2 (Ω)	504	625.3
C_3 (pF)	5.35	4.36
Yield Estimate	29.3%	77.5%

To gain insight into the statistical properties of the minimax solution, yield factor histograms for the variables Z, R1, R2, and C3 are shown in Figures 7.9(a through d), respectively. From these yield factor histograms we see that the yield is sensitive to the values of Z, R1 and R2, and the value of R2 probably needs to be raised to increase yield. Remember that these results are shown for a single-point optimized (minimax) design. This minimax single-point optimized solution shows a classical brinksmanship behavior. From the slopes on the YFHs it is apparent that the yield can be improved by statistical optimization. Figure 7.9(a) shows that the yield is sensitive to the value of Z, yet we will not optimize on Z during yield optimization.

Following the procedure from [4], we used R1, R2, and C3 as optimization variables, while the variables listed in Table 7.1 are varied according to their statistical model. A center-of-gravity type of statistical optimization was applied to this circuit. The final results along with the initial values for the optimization variables are given in Table 7.2.

7.3.5 Results

The data in Table 7.2 show two important results. First, the yield is improved to approximately 77% with the application of statistical optimization on only three variables. The FET parameters, although statistically varying in this example, were not changed by the statistical optimization. Second, the change made to component R2 is relatively large. Sometimes large changes in parameter values are needed to optimize the yield. However in other circuits, small changes in nominal component values are all that is needed. The large change in C3 is likely due to the yield being relatively insensitive to the value of C3 and any of a large range of values would work well.

The YFHs for Z, R1, R2, and C3 for the statistically optimized design are shown in Figure 7.10. Notice that the allowed variation in Z for the statistically optimized design (Figure 7.10a) is wider than the allowed variation for Z in the single-point optimized design (Figure 7.9a). The statistical optimizer increased the value of R2 to allow for a wider range of acceptable Z values.

As a final aid in seeing the effect of yield optimization on this circuit, statistical response trace plots before and after yield optimization are shown in Figure 7.11(a, b) respectively. As seen in these figures the effect of the statistical optimization is to raise the ensemble of the [S21] curves and VSWR curves. Sometimes a compressing of the variation is observed in the statistical trace plots before and after statistical optimization, but this is not evident here.

7.4 EXAMPLE—A SATELLITE-COMMUNICATIONS SYSTEM

7.4.1 Comments

This example comes from one of the few published works showing the benefits of statistical design when applied to the design of systems. A good method for sta-

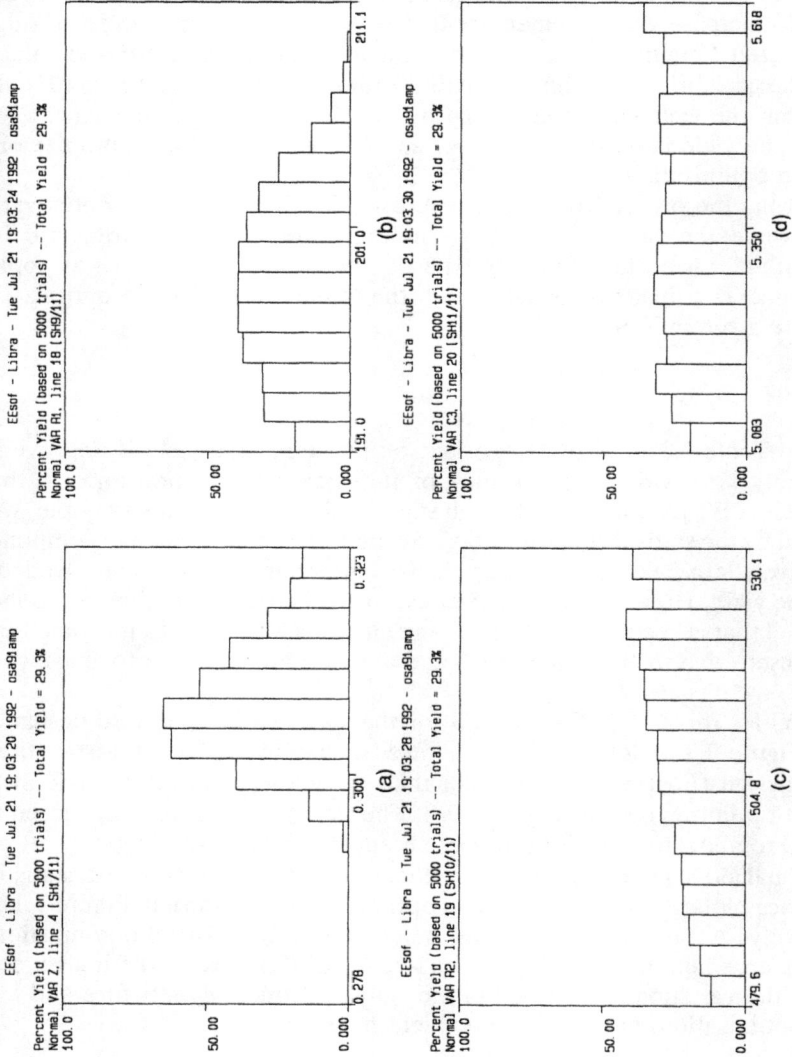

Figure 7.9 Yield factor histograms of (a) variable Z, (b) variable R1, (c) variable R2 and (d) variable C1, for the minimax single-point optimized design.

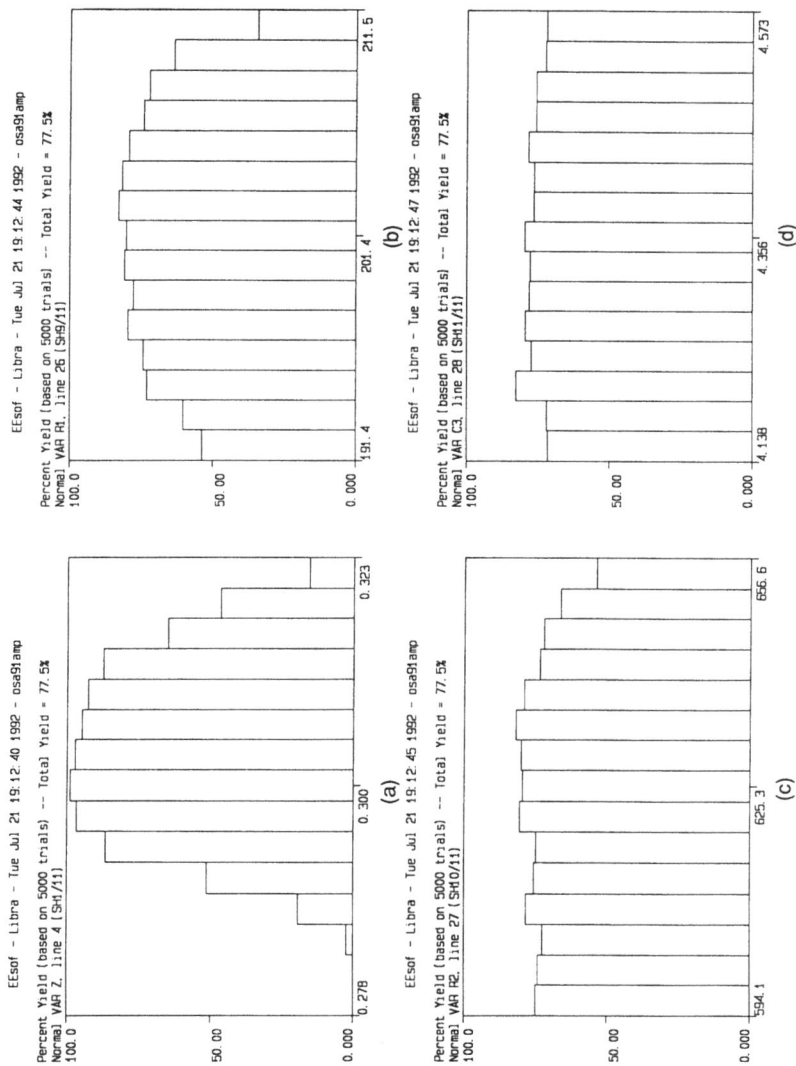

Figure 7.10 Yield factor histograms on (a) variable Z, (b) variable R1, (c) variable R2 and (d) variable C1, for the statistically optimized design.

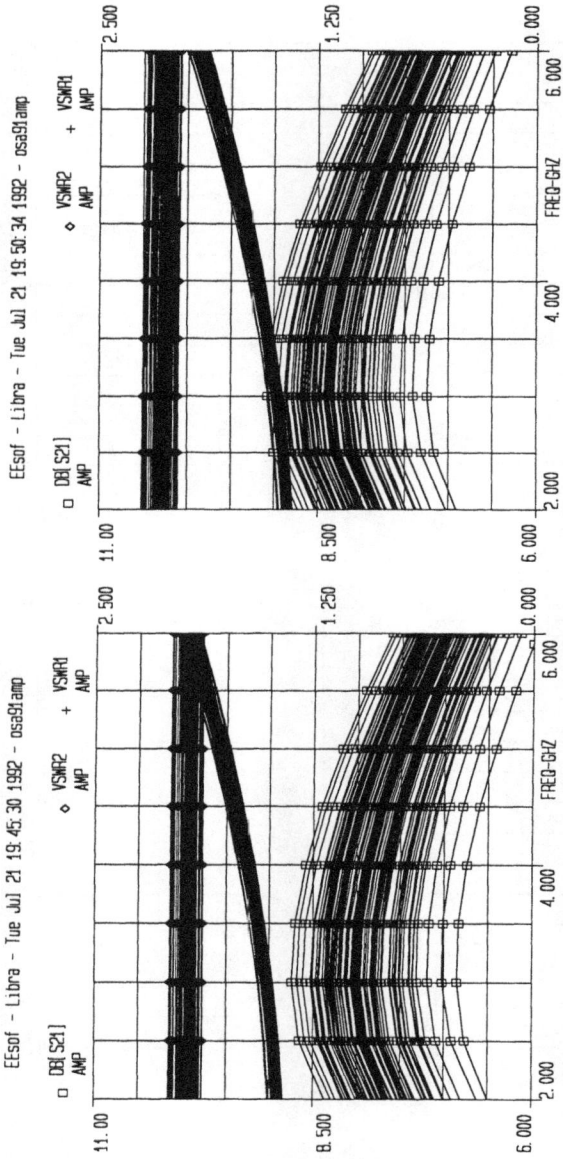

Figure 7.11 Statistical trace plot of S21, VSWR1 and VSWR2 (a) before, and (b) after statistical optimization.

tistical design that applies to any application is also presented. The example is a bit ideal but hopefully it communicates that there are no real application limitations to statistical-design methods. This work was accomplished by Roland Cooke.

7.4.2. Preliminary Information

In the last year, commercial statistical circuit-design software (i.e., design for high manufacturing yield) has been introduced to the microwave community [5, 6]. This example presents the application of state-of-the-art statistical-design techniques to systems rather than circuits. We use the CAD systems package OMNISYS [7] for this book, however these techniques can be applied by modifying any systems-level simulator.

Typically systems are designed to exceed the specified performance. This gives the design a performance margin to handle the inevitable performance degradation during manufacture or the system's lifetime operation. For instance, expensive linear amplifiers are used to insure harmonic-distortion specifications are met. Any performance margin can be costly if not properly managed and analyzed. In most cases it is felt that application of statistical-design techniques will allow the system designer to relax the individual element specifications in systems design, and still meet overall performance and yield specifications for the system. This will result in lower system cost, with no sacrifice in the systems specifications or reliability.

7.4.3 Statistical System-Design Methodology

System statistical optimization (SSTATO) is accomplished following these steps [8]:

1. Model the statistical parameters.
2. Specify the system performance test.
3. Analyze the system.
4. Determine tolerance and sensitivity.
5. Optimize by design centering.

These steps form a basic methodology for statistical design.

Step 1—Model the Statistical Parameters

Determine all of the parameters of interest and ascribe statistical information to them. An example of this is an amplifier. Besides gain, it also has input and output reflection coefficients, third-order intercepts, and 1-dB compression points that describe its linearity. Nominal values (i.e., average values) as well as statistical distributions (i.e., Gaussian and Uniform) and correlations must be assigned to each system parameter. High-order statistical models may be necessary for accurate design and analysis.

Step 2—Specify the System Performance Test

A criterion for entire-system acceptance must be established. This is done by determining the system-performance specification, which depends on the system's application. For example, it might be necessary to achieve a certain output signal-to-noise ratio, or perhaps group delay, or power. These constraints on the system set the acceptance and rejection criterion for yield analysis.

Step 3—Analyze the System

Once all of this statistical information is determined and the desired performance is set, the system component parameters should be exercised to the full extent of their statistical variation, and the system yield determined. This can be done by Monte Carlo yield analysis.

Step 4—Determine Tolerance and Sensitivity

The system performance can then be analyzed as a function of parameter variation, referenced to yield using the yield factor histogram. The system parameter variation can then be tightened (tolerances reduced), moved (design-centered), or left alone (if variation has no effect). This step is reiterated as many times as necessary to come close to the desired yield and system performance. This manual step is recommended to both improve the designers "feel" for the system and speed the design process.

Step 5—Optimize by Design Centering

Finally the optimal parameter nominal values are determined by use of statistical-analysis and design software, hence optimizing system yield (design centering).

7.4.4 Analogue—Two Amplifiers and a Filter [8]

This simple example helps to introduce the procedure for system statistical analysis and design. The two amplifiers have gain A1 and A2 (in dB) respectively and the gains are independent and vary uniformly ±10%. The filter bandwidth is also a statistical variable which varies ±90% or ±50%. IP3 is the third-order intercept point of the amplifier (in dBm) and 1DBC is the 1-dB compression point in dBm. The input drive level for each example is −70 dBm.

The performance specifications for this example are chosen as signal-to-noise ratio at the filter output (s/n) in dB; group delay at the output (gd) in μs; and power out (pw) in dBm. These were chosen because good group delay requires a wide bandwidth, while good s/n requires a narrow bandwidth. Also, good s/n requires linear amplifiers to avoid harmonic distortion and subsequent loss of carrier power.

This system is shown in Figure 7.12. For this example, the second amplifier's IP3 was set so that it is operating in its nonlinear region. Specifications and results of yield calculations are shown in Table 7.3. Nominal parameter values and changes due to design centering are found in Table 7.4. Typical yield factor histograms are shown in Figure 7.13.

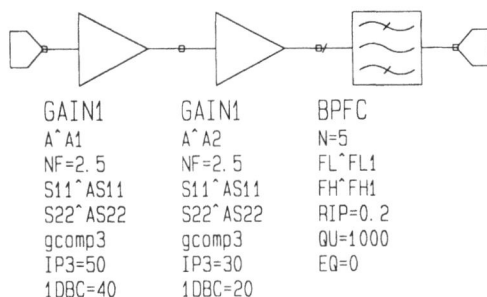

```
GAIN1          GAIN1          BPFC
A^A1           A^A2           N=5
NF=2. 5        NF=2. 5        FL^FL1
S11^AS11       S11^AS11       FH^FH1
S22^AS22       S22^AS22       RIP=0. 2
gcomp3         gcomp3         QU=1000
IP3=50         IP3=30         EQ=0
1DBC=40        1DBC=20
```

Figure 7.12 Nonlinear-system block diagram.

Table 7.3
Nonlinear Amplifiers—Confidence Intervals (1,000 trials) ±3%

ampl varies	a1 = 72 ± 10%	a1 = 72 ± 10%	a1 = 72 ± 10%
amp2 varies	a2 = 30 ± 10%	a2 = 30 ± 10%	a2 = 30 ± 10%
BW varies	BW = 1mhz ± 90%	BW = 1mhz ± 50%	BW = 1mhz ± 50%
spec's	s/n>10 gd<.05	s/n>10 gd<.05	s/n>10 gd<.05
			pw>16
s/n only	40.9%	35.5%	35.5%
gd only	77.3%	99.9%	99.9%
pw only	no spec	no spec	80.0%
all spec'd	20.9%	33.5%	22.3%
design center	no data	no data	32.0%

Table 7.4
Nominal Parameter Values and Change.

	Nominal	Design centered	Change
a1	72	72	0
a2	30	30.5	+0.5db
BW1	0.1	0.081	0.019GHz

Linear Amplifier

For the linear-amplifier case the system yield is not dependent on the gains of the amplifiers. The yield for the linear-amplifier case therefore goes up to 73.8% uncentered and 80.6% design-centered.

7.4.5 Analogue—A Satellite Receiver [8]

The second example involves analysis of a 4.0-GHz satellite-downlink-receiver system with similar specifications and parameter variations as in the above example. A simplified system block diagram is shown in Figure 7.14. The exact block diagram of the system is in the OMNISYS(TM) applications manual [9].

Four statistical parameters were chosen for this system. They are two amplifier gains (A1, A2), a mixer local oscillator frequency (LO1), and finally the system output filter bandwidth (BW1). Otherwise the system was used as given. After system component parameter tolerance and sensitivity changes were made the yield was improved from 23.4% to 100%. The improvement due to optimization is dependent upon the starting design. In this analysis the dramatic yield improvement results come from the system local oscillator frequency being relocated. This was done to test the simulator's ability to optimally reposition the frequency, and to demonstrate more clearly design-centering results. The histograms before design centering are shown in Figure 7.15.

Specifications and results of yield analysis are shown in Table 7.5. Nominal parameter values and change are found in Table 7.6.

7.4.6 Summation

The application of established circuit statistical-analysis and design techniques to systems can be very beneficial. These tools applied to systems design allow the systems engineer to perform statistical-sensitivity studies to determine sensitivities and tolerances of the important systems components. Robust and reliable systems designs can result from the application of these techniques. Future applications include antennas, phased-array radar, and cellular communications, to name a few.

EEsof - OmniSys - Tue Dec 4 15: 26: 12 1990 - athesis2b. syd

Percent Yield (based on 1000 trials) -- Total Yield = 35.5%
Uniform Parameter A1 of 72.000 at nodes on line #14 (SH1/3).

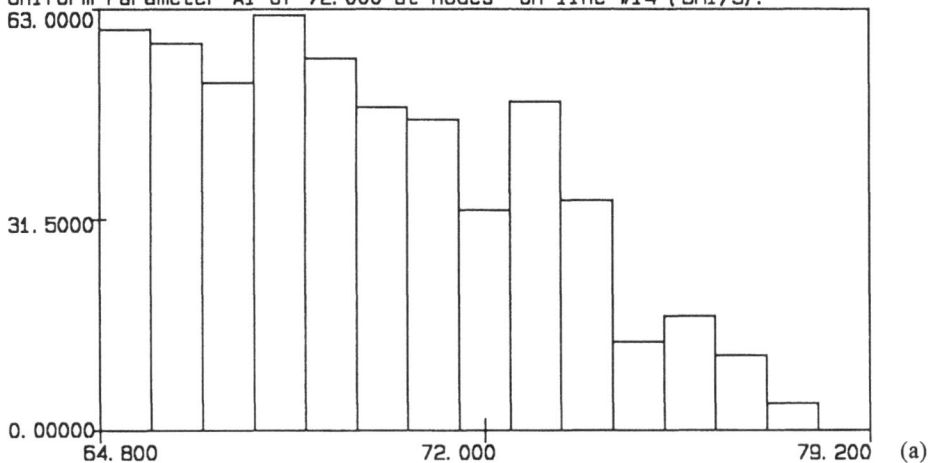

(a)

EEsof - OmniSys - Thu Dec 6 14: 57: 02 1990 - athesis2c. syd

Percent Yield (based on 1000 trials) -- Total Yield = 22.3%
Uniform Parameter A1 of 72.000 at nodes on line #15 (SH1/3).

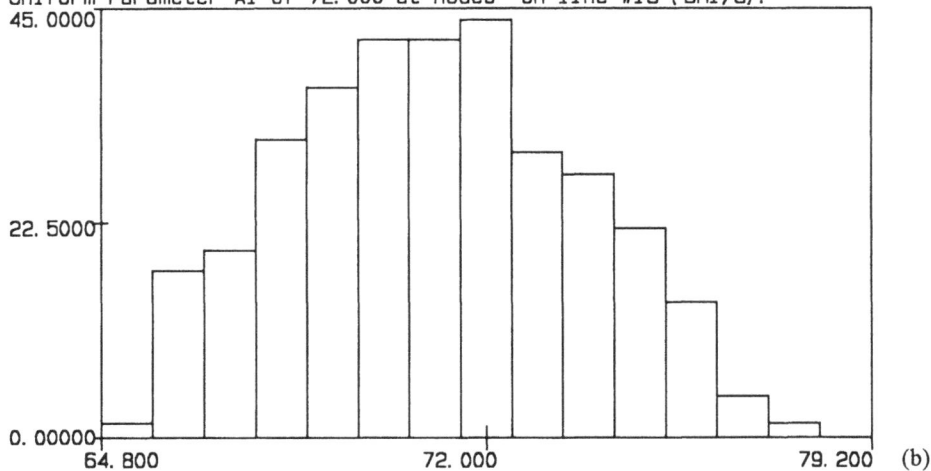

(b)

Figure 7.13 Yield factor histograms of a nonlinear system: (a) the effect of A1 on yield with the signal-to-noise ratio as the system specification (the nonlinearity of amplifier 2 causes this surprising result); (b) the effect of A1 on yield with signal-to-noise ratio, group delay and power out specified.

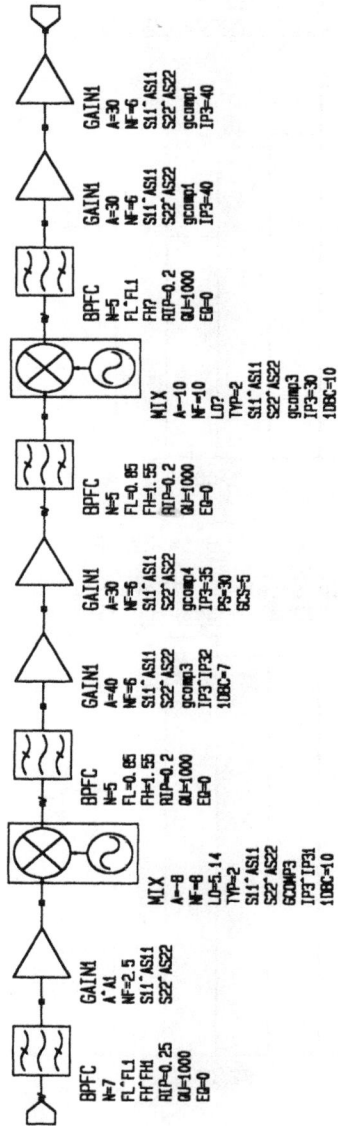

Figure 7.14 4.0-GHz satellite-receiver block diagram.

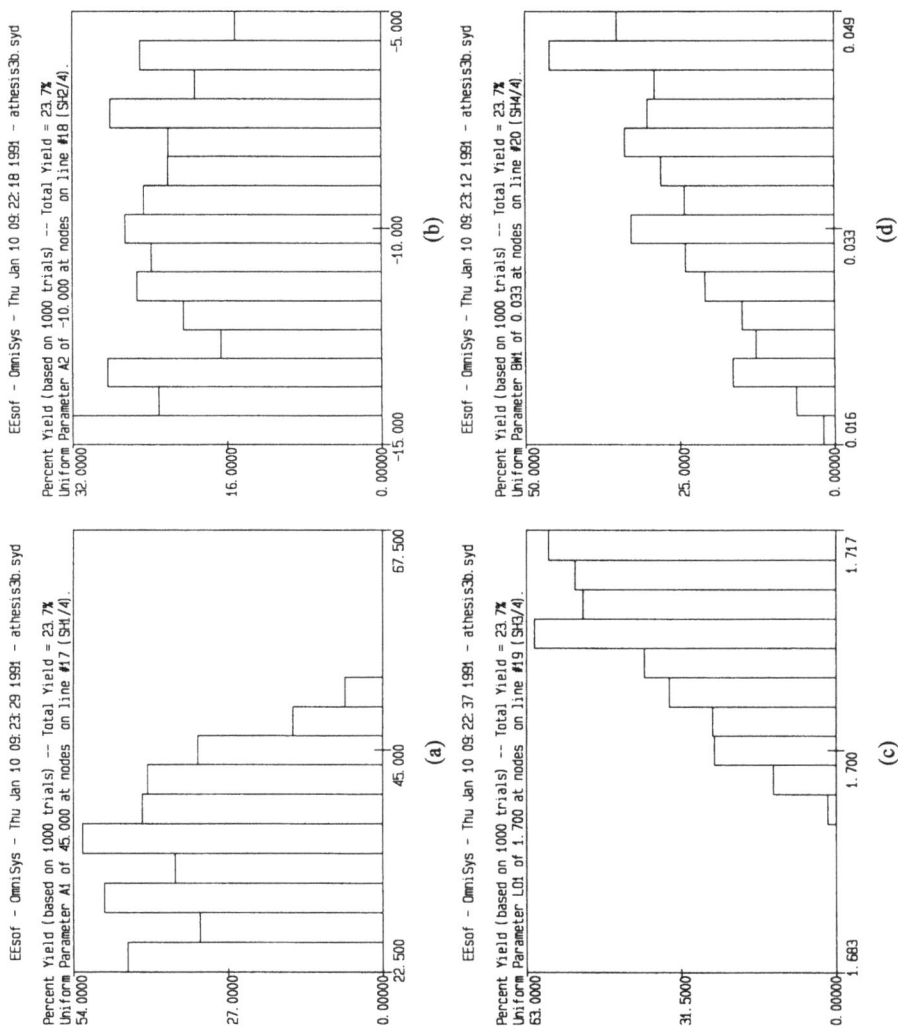

Figure 7.15 (a) A1, (b) A2, (c) LO1, and (d) BW1 before design centering.

Table 7.5
Satellite Receiver—95% Confidence Intervals (1000 trials) < 2%

amp1	a1 = 45 ± 50%	a1 = 45 ± 10%
amp2	a2 = − 10 ± 50%	a2 = − 10 ± 10%
local osc.	lo1 = 1.70 ± 1%	lo1 = 1.70 ± 0.25%
bandwidth	BW1 = 33 ± 50%	BW1 = 33 ± 10%
spec's	s/n>11 gd<0.45	s/n>11 gd<0.45
s/n	26.4%	no data
gd	91.3%	no data
s/n and gd	23.7%	20.9%
design center	74.4%	100%

Table 7.6
Nominal Parameter Values and Change

	Nominal	Design centered	Change
a1	45	38.24	− 6.76dB
a2	− 10	− 9.93	+ 0.07db
lo1	1.70	1.71	+ 0.01GHz
BW1	0.033	0.037	+ 0.004GHz

7.5 CASE STUDY—A 0.5- TO 2.5-GHz MMIC GAIN BLOCK

7.5.1 Comments

In this case study we examine the power and accuracy of the empirical-database model (truth model) by comparing the predicted and manufactured statistical response of a GaAs MMIC 0.5- to 2.5-GHz amplifier. These results were first reported in [10] and were developed by T. Wandinger and D.A. Fisher at Watkins-Johnson Company.

7.5.2 Preliminary Information

A 0.5- to 2.5-GHz feedback amplifier [11] has been designed using two 0.5- by 300-μm FETs. The amplifier is intended to be a drop-in replacement for an existing hybrid amplifier and, as such, requires excellent gain flatness and low DC bias current. The single-stage gain needs to be greater than 13.5 dB and a low-noise figure is not required. To ensure ease of integration at the system level, the VSWR must be better than 2:1 out to 6 GHz. To reduce assembly costs and verify the circuit performance on-wafer, all bias circuitry was included on the chip. Figure 7.16 shows the schematic diagram of the MMIC amplifier.

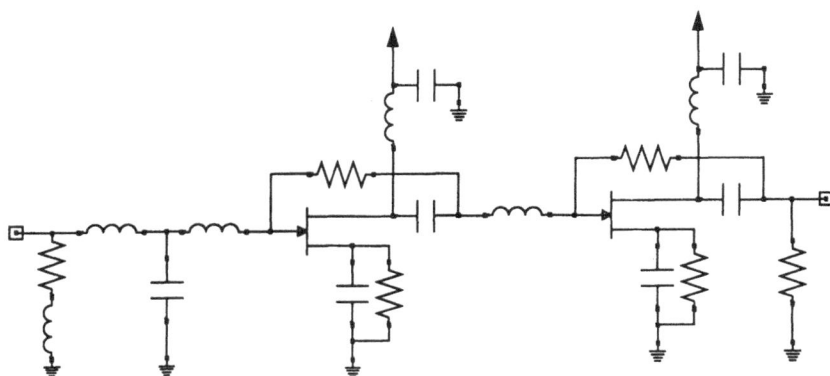

Figure 7.16 Schematic diagram of the 0.5- to 2.5-GHz MMIC Amplifier.

7.5.3 Amplifier Design

Circuit Design

A feedback topology was chosen in order to provide flat gain over the wide bandwidth. The gate periphery was optimized to deliver the necessary output power at the required bias condition. The small signal FET performance was calculated from an equivalent circuit model. The model's values were derived from on-wafer measured S-parameters of test FETs. The S-parameters are measured for every wafer processed at Watkins-Johnson Company (W-J) and a database is used to arrive at a statistically valid model. A lossy match is used on both the input and output to achieve good VSWR over the complete frequency range. The spirals were taken from a library of characterized elements and are represented in the circuit design by their measured S-parameters.

Circuit Fabrication

The amplifier chips are fabricated using W-J's 0.5-μm MMIC process. Ion-implanted wafers are used with a peak carrier concentration in the channel layer of 4.0 by 1017 cm^{-3}. Device isolation is achieved using an oxygen implant. Ohmic contacts are formed from an Ni/Au/Ge metalization. Silicon nitride serves as both the capacitor dielectric and the chip passivation. Precise resistor values are realized using TiWN thin-film resistors. Electroplating of air bridges and transmission lines completes the front-side processing. The substrates are lapped to a thickness of

110 μm and vias are laser-drilled from the back side. Following chip separation the fabrication is complete. The final chip size is 2.1 by 1.5 by 0.1 mm. Figure 7.17 shows a picture of the MMIC.

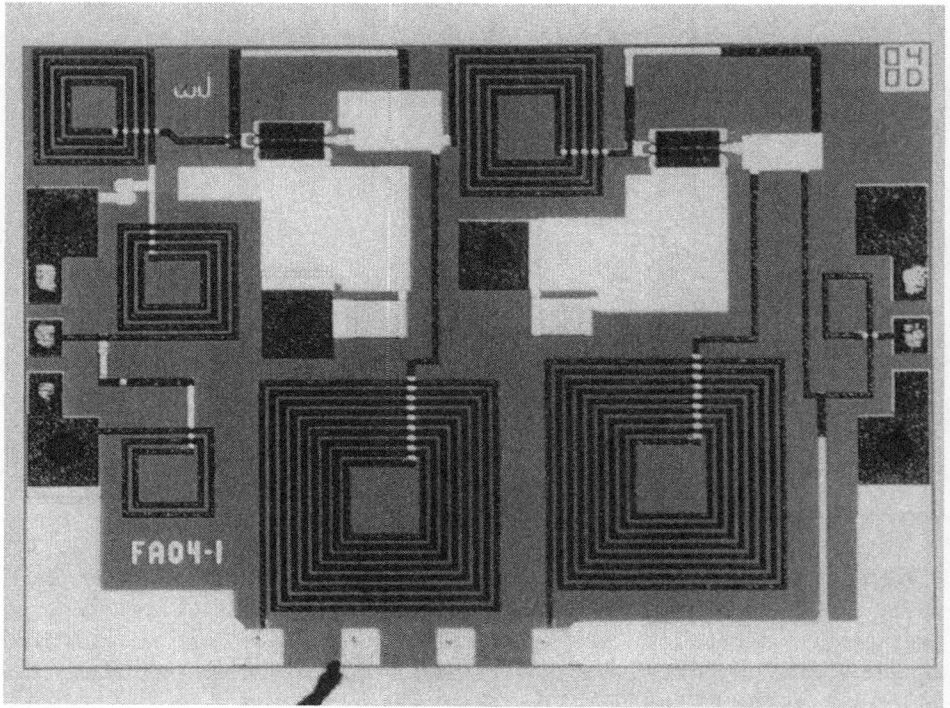

Figure 7.17 Photograph of the 0.5- to 2.5-GHz MMIC amplifier.

7.5.4 Statistical-Response Prediction With the Database Model

Circuit Statistical-Response Performance

Turn-on performance showed the amplifier to be a first-pass design success. The measured and simulated statistical-response plots are shown in Figures 7.18 and 7.19 respectively. A nominal gain of 14.5 ± 0.2 dB from 0.5 to 2.5 GHz was measured using the Cascade Microtech wafer probe. The input and output VSWR was measured to be better than 1.7:1 across the band. Except for the very low end of the band, the input VSWR is typically better than 1.35:1 and the output VSWR better than 1.25:1. These numbers are averages from the first two lots of wafers. The amplifier's performance has been reproducible from wafer to wafer and wafer

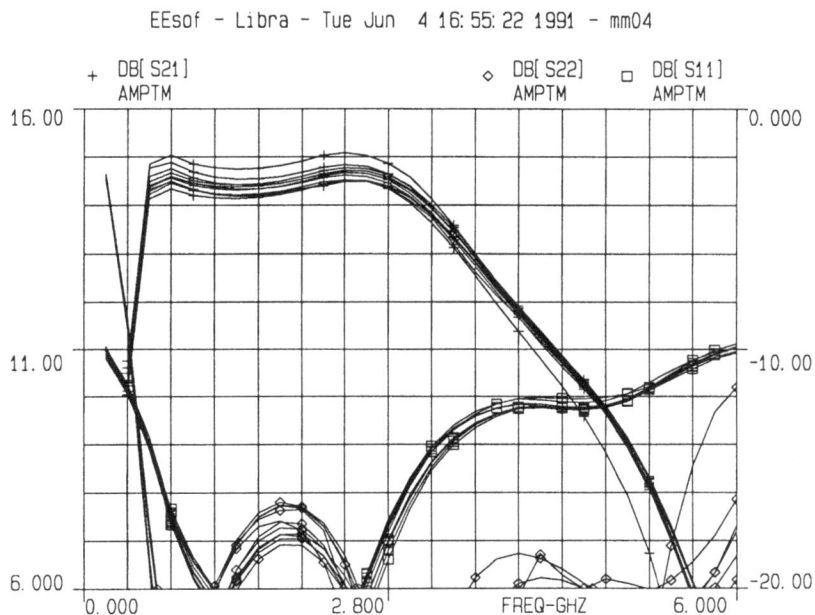

Figure 7.18 Measured statistical response of the 0.5- to 2.5-GHz MMIC amplifier.

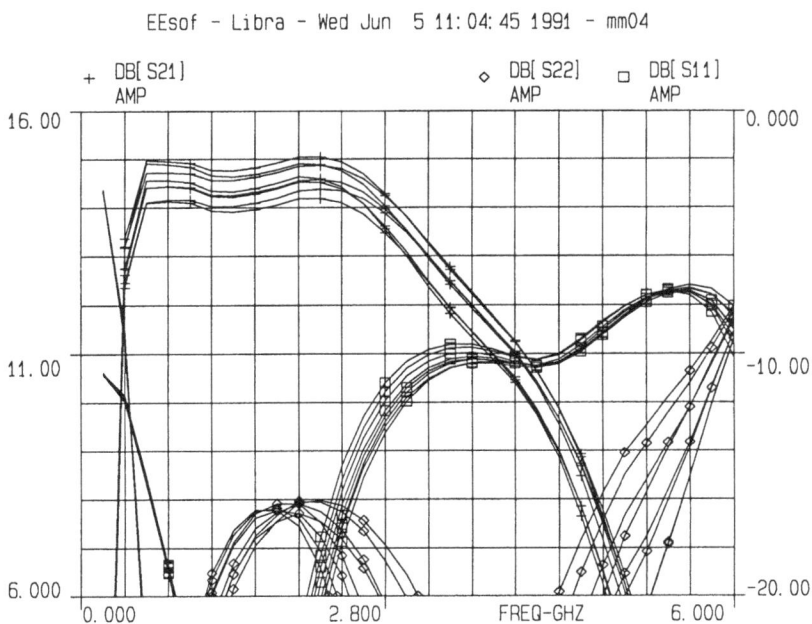

Figure 7.19 Simulated statistical response of the 0.5- to 2.5-GHz MMIC amplifier.

lot to wafer lot. Chips measured in a test fixture have shown excellent correlation to the on-wafer RF data. The amplifier uses less than 80 mA of bias current at 5V.

7.5.5 Summation

A demonstration of the use of a database parameter statistical model for GaAs FET parameters is shown. This example illustrates the excellent predictive power of a database model by showing close agreement between the measured and simulated statistical response of a 0.5- to 2.5-GHz GaAs MMIC amplifier. Application of database models is not restricted to device type or even for active devices. In any situation where complex high-order statistical relationships exist, database models should be useful.

7.6 CASE STUDY—SMALL-SIGNAL YIELD ANALYSIS

7.6.1 Comments

This case study describes a technique for computing small-signal RF yield of monolithic microwave integrated circuits based on the calculated single-point sensitivity of electrical-model parameters to physical-device parameters. As shown in Chapter 4, single-point sensitivities must be used with caution in statistical analysis. However the results of this work are good. The RF yield-analysis approach used here uses key MMIC-device physical variables in conjunction with the computed sensitivities of electrical-model parameters as the statistical-parameter model. This work was contributed by D.L. Allen, and was first reported by D.L. Allen, J. Beall, and M. King of Texas Instruments [12].

7.6.2 Preliminary Information

In previous work, the statistical design of microwave circuits has focused on the determination of parametric variation of the GaAs FET devices only. FET equivalent circuit model parameter statistics were studied by Purviance et al. [13]. This work demonstrated that low-order statistical data on the FET parameters is not sufficient to characterize the statistics on S-parameters. Instead Purviance et al. proposed the use of the truth model which is based on performing a statistical analysis using the original FET database. Campbell [14] improved the truth model by interpolating between data points in the FET database. These methods are accurate but have the disadvantage of requiring a large database for each separate FET.

In this case study, we use a statistical model for GaAs MMICs that is based on the single-point sensitivity of the electrical-model parameters to physical-device parameters involved in the processing of MMIC circuits. This method is flexible and is easily applied to FETs and other MMIC-circuit elements.

There are several advantages to this approach that include:

1. The approach allows the use of a statistically independent variable set (although statistical independence is not verified in this work).
2. RF yield is easily estimated for different device profiles.
3. Passive and active MMIC-circuit parameters are included in the analysis.
4. Circuit variability can be traced to process control points.
5. The approach makes use of a large database of device physical measurements which are made on every wafer.
6. The approach can be implemented using almost any nominal FET model.

7.6.3 Approach

A block diagram of the yield-estimation process is shown in Figure 7.20. The approach consists of combining the statistical variation of physical-device parameters and sensitivity equations for each microstrip or FET device electrical-model parameter with a linear circuit file describing a given MMIC device. The sensitivity equations relate the key electrical parameters of the MMIC circuit to the physical-process variables. In this work, the physical-device and microstrip parameters as shown in Table 7.7 are included in the VAR block of a Touchstone/Libra® circuit file while the device and microstrip sensitivity equations are described in the EQN block. The data for the parameters included in these blocks consists of the parameter mean value and standard deviations which are routinely derived from GaAs foundry data on each wafer. Additionally, a yield specification for the MMIC

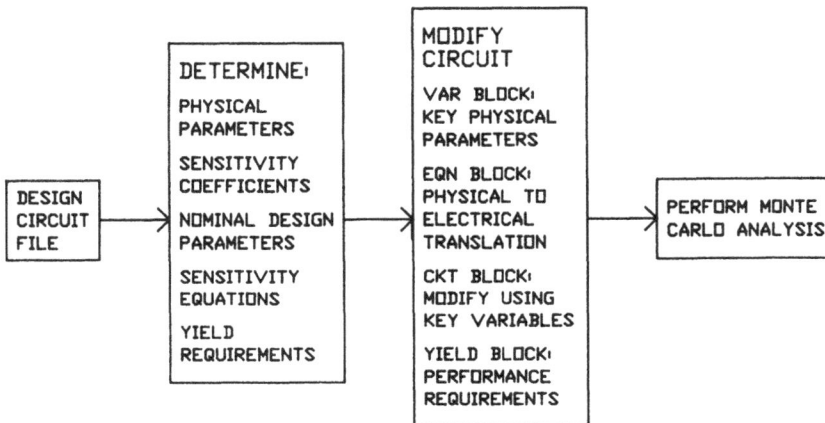

Figure 7.20 Yield-prediction process.

Table 7.7
Physical-MMIC-Device Variables

	Parameter	*Symbol*	*Std. Dev.*
Linewidth	Isolation	DLWI	0.140 μm
Parameters	Ohmic	DLWOH	0.314 μm
	TaN	DLWTA	0.377 μm
	TaN Contact	DLWTA	1.767 μm
	TF Resistor	DLWTF	1.312 μm
	First Metal	DLWFM	0.475 μm
	Cap Top Plate	DLWTP	0.475 μm
	Plated Metal	DLWPM	0.700 μm
GaAs Resistors	Activation	DA	0.100
	Surface Potential	DPHIS	0.490
	Contact Resistance	DRC	0.45 Ω − mm
MIM Capacitors	Cap Top Plate	DLWTP	0.475 μm
	Capacitance/Area	CMIM	7pF/mm^2
FETs	Gate Length	DLG	0.0389
	Narrow Recess	DLNR	0.052
	Narrow Recess	DISAT	0.059
	Wide Recess	DLWR	0.093
	Wide Recess	DISAT	0.037
	Source-Drain	DLSD	0.047
	Gate-Drain	DLGS	0.054
	Isolation	DLWI	0.140 μm
	Activation	DA	0.100
	Surface Potential	DPHIS	0.490
	Recessed Potential	DPHISR	0.110
	Contact Resistance	DRC	0.45 Ω − mm

circuit is specified in the YIELD block. Once the VAR, EQN, and YIELD blocks are defined, the Monte Carlo analysis features of Touchstone/Libra are used to compute the MMIC-device yield. During the Monte Carlo analysis one can either compute yield factor histograms, or compute a histogram of performance to determine a performance mean and standard deviation.

7.6.4 Sensitivity Equations and Coefficients

In the analysis, the relationship of device electrical-model parameters to the physical parameters is described by a set of single-point sensitivity equations which are first-order n-dimensional Taylor-series approximations. As an example, the sensitivity equation for C_{gs} is shown below.

$$C_{gs} = C_{gs0}*\left(1 + S_{WI}^{C_{gs}}*\left(\frac{DWLI}{WGF}\right) + S_{LG}^{C_{gs}}*DLG + S_{LSD}^{C_{gs}}*DLSD\right)$$
$$+ C_{gs0}*(S_A^{C_{gs}}*DA + S_{ISAT}^{C_{gs}}*DISAT + S_{PHIS}^{C_{gs}}*DPHISR)$$

In this equation, C_{gs0} is the nominal or design value. The other terms consist of sensitivity coefficients (i.e. $S_{LG}^{C_{gs}}$) and physical-parameter percent standard deviations. Standard deviations for the physical variables considered are shown in Table 7.7. During yield analysis, the parameter C_{gs} will vary randomly as a function of the physical variables shown. MMIC circuits with different device profiles can easily be analyzed by changing the appropriate sensitivity coefficients. Other intrinsic FET parameters are described in a similar manner using the equations in Table 7.8.

Table 7.8
FET Sensitivity Equations

$$C_{gs} = C_{gs0}*\left(1 + S_{WI}^{C_{gs}}*\left(\frac{DWLI}{WGF}\right) + S_{LG}^{C_{gs}}*DLG + S_{LSD}^{C_{gs}}*DLSD\right)$$
$$+ C_{gs0}*(S_A^{C_{gs}}*DA + S_{ISAT}^{C_{gs}}*DISAT + S_{PHIS}^{C_{gs}}*DPHISR)$$

$$R_g = R_{g0}*(1 + S_{LG}^{R_g}*DLG)$$

$$C_{gd} = C_{gd0}*\left(1 + S_{WI}^{C_{gd}}*\left(\frac{DLWI}{WGF}\right) + S_{LG}^{C_{gd}}*DLG + S_{LSD}^{C_{gd}}*DLSD + S_{LGS}^{C_{gd}}*DLGS\right)$$

$$gm = gm_0*\left(1 + S_{WI}^{gm}*\left(\frac{DLWI}{WGF}\right) + S_A^{gm}*DA + S_{ISAT}^{gm}*DISAT + S_{PHIS}^{gm}*DPHISR\right)$$

$$T = T_0*(1 + S_{LG}^{T}*DLG)$$

$$R_s = R_{s0}*\left(1 + S_{WI}^{R_s}*\left(\frac{DLWI}{WGF}\right) + S_{LG}^{R_s}*DLG + S_{LNR}^{R_s}*DLNR + S_{LWR}^{R_s}*DLWR + S_{LSD}^{R_s}*DLSD\right)$$
$$+ R_{s0}*(S_A^{R_s}*DA + S_{ISAT}^{R_s}*DISAT + S_{ISATW}^{R_s}*DISATW + S_{PHIS}^{R_s}*DPHIS + S_{DRC}^{R_s} + S_{LGS}^{R_s}*DLGS)$$

$$C_{ds} = C_{ds0}*\left(1 + S_{WI}^{C_{ds}}*\left(\frac{DLWI}{WGF}\right) + S_{LSD}^{C_{ds}}*DLSD\right)$$

$$R_{ds} = R_{ds0}*\left(1 + S_{WI}^{R_{ds}}*\left(\frac{DLWI}{WGF}\right)\right)$$

$$R_d = R_{d0}\left(1 + S_{WI}^{R_d}*\left(\frac{DLWI}{WGF}\right) + S_{LG}^{R_d}*DLG + S_{LNR}^{R_d}*DLR + S_{LWR}^{R_d}*DLWR + S_{LSD}^{R_d}*DLSD\right)$$
$$+ R_{d0}*(S_A^{R_d}*DA + S_{ISAT}^{R_d}*DISAT + S_{ISATW}^{R_d}*DISATW + S_{PHIS}^{R_d}*DPHIS + S_{RC}^{R_d}*DRC + S_{LGS}^{R_d}*DLG$$

where

$$S_\alpha^\beta = \frac{\alpha_0}{\beta_0}\frac{\partial\beta}{\partial\alpha}$$

β = electrical parameter
α = physical parameter

The variation in MMIC performance due to other circuit elements such as MIM capacitors and GaAs resistors can also be computed. The sensitivity equations for these two circuit elements are given below.

$$C = C_{mim}*(WO + DLWTP)*(LO + DLWTP)$$

where C_{mim} = capacitance/unit area; and

$$R = R_{s_0}*\left(1 + S_A^{R_s}*DA + (S_{PHIS}^{R_s}*DPHIS)\frac{(LO - DLWOH)}{(WO - DLWOH)} + \frac{2*R_c(1 + DRC)}{WO + DLWISO}\right)$$

where R_{s0} the GaAs sheet resistance, and R_c is the contact resistance.

A variety of microstrip-circuit elements can be described by similar sensitivity equations and are limited only by the availability of a suitable model and the total number of variables and equations needed to describe a MMIC circuit. Circuit elements included in the RF yield model to date have included FETs, metal-insulator-metal (MIM) capacitors, thin-film resistors, critical transmission lines, open-circuit stubs, and substrate parameters.

To perform a yield analysis, the single-point sensitivity of electrical model parameters (i.e., the sensitivity coefficients) as a function of physical process parameters is needed. The sensitivity coefficients for the intrinsic FET parameters C_{gs}, C_{gd}, C_{ds}, g_m, and sheet resistance were computed using a device-physics-based program developed at Texas Instruments called FETMOD. For the model parameters R_{ds}, R_d, R_s, and R_g, the sensitivity coefficients were computed manually using data derived from the GaAs foundry database. An example of the sensitivity coefficients for a low-noise high-current FET profile is shown in Table 7.9. Nonzero values in this table represent the primary sensitivities of an intrinsic FET parameter to a given physical parameter.

7.6.5 Analogue—A Broadband Low-Noise MMIC Distributed Amplifier

To demonstrate the accuracy of the proposed approach, the following example, an RF yield analysis of a broadband low-noise MMIC distributed amplifier is described and compared to measured data. The low-noise amplifier shown in Figure 7.21 has a total gate periphery of 756μm and is comprised of four single-gate FETs with gate widths of 189μm each. This MMIC amplifier was fabricated using TI standard (low-noise high-current) ion-implant material (peak doping approximately 1.0×1018 cm^{-3}). The physical-parameter standard deviations and sensitivity coefficients for this material type are given in Tables 7.7 and 7.8. The RF

Table 7.9
Low-Noise High-Current Profile Sensitivity Coefficients

	DLW1	DLG	DLNR	DLWR	DLSP	DA	DISAT	DISATW	DPHIS	DPHISR	DLGS
R_g	0.0	-1.18	0.0	0.0	0.0	0.0	0.0	0.0	0.0	0.0	0.0
C_{gs}	1.0	0.81	0.0	0.0	-0.02	0.24	-0.54	0.0	-0.25	0.0	0.0
R_i	0.0	0.0	0.0	0.0	0.0	0.0	0.0	0.0	0.0	0.0	0.0
C_{gd}	1.0	0.28	0.0	0.0	-0.25	0.0	0.0	0.0	0.0	0.0	0.24
C_i	0.0	0.0	0.0	0.0	0.0	0.0	0.0	0.0	0.0	0.0	0.0
g_m	1.0	0.0	0.0	0.0	0.0	0.30	-0.68	0.0	-0.32	0.0	0.0
T	0.0	1.0	0.0	0.0	0.0	0.0	0.0	0.0	0.0	0.0	0.0
R_s	-1.0	-0.08	0.09	0.12	0.58	-0.19	-0.03	-0.20	0.07	0.12	0.54
C_{ds}	1.0	0.0	0.0	0.0	-0.22	0.0	0.0	0.0	0.0	0.0	0.0
R_{ds}	-1.0	0.0	0.0	0.0	0.0	0.0	0.0	0.0	0.0	0.0	0.0
R_d	-1.0	-0.07	0.09	0.11	0.54	-0.18	-0.03	-0.18	0.06	0.11	-0.50

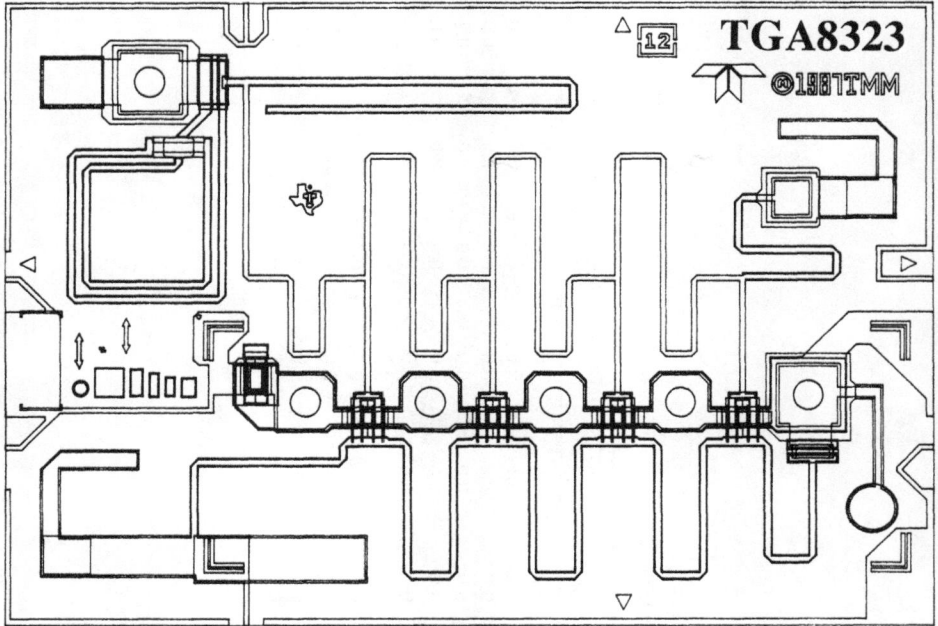

Figure 7.21 Broadband low-noise distributed amplifier.

RF yield model for this MMIC amplifier includes models of the FETs, MIM capacitors, GaAs resistors, and inductive-bond wires.

After modifying the design circuit file and adding the physical-variable standard deviations, sensitivity equations, and yield requirements, a Monte Carlo analysis was performed on the low-noise amplifier. The predicted versus measured statistical variations in small-signal gain are shown in Figures 7.22 and 7.23. For the computed performance, 100 trial sweeps were used while the measured data was derived from approximately 200 devices. To compute an accurate mean and standard deviation, a performance histogram of gain (Figure 7.24) using 5000 trials was computed at 10 GHz. In summary the predicted mean and standard deviation in gain at 10 GHz is 8.08 dB and 0.44 dB and the measured mean and standard deviation was 8.01 dB and 0.54 dB. The predicted standard deviation is 0.1 dB less than the measured and is attributable to variations in critical transmission lines and variations in gate length which impact the device transconductance. The sensitivity of gm with respect to gate length has not been included in the model to date. Overall, there is an excellent agreement in the predicted versus measured mean and standard deviation in gain versus frequency. Good agreement was also

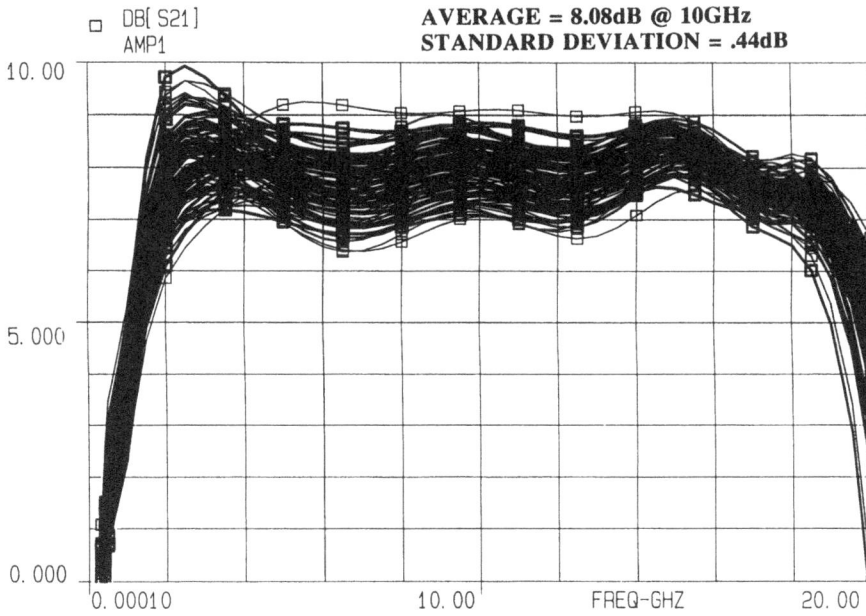

Figure 7.22 Predicted small-signal gain variation.

achieved between the predicted versus measured performance for other parameters such as the input return loss as shown in Figures 7.25 and 7.26. Using a yield requirement of 6.5dB < G < 9.0dB for a 2- to 18-GHz bandwidth, the RF yield of this low-noise amplifier was predicted at 80%.

7.6.6 Summation

In summary a technique for computing predicted small-signal RF yield of a MMIC-circuit design based on physical parameters has been described. This technique uses commonly measured FET physical parameters in conjunction with electrical-parameter sensitivity equations to model both passive and active circuit elements. A linear microwave-circuit simulator is then used to compute the RF yield through Monte Carlo analysis. This RF yield-analysis method has successfully been applied to several MMIC amplifiers with good results between measured and computed performance. Future work is in progress to adapt this technique to large-signal microwave circuits.

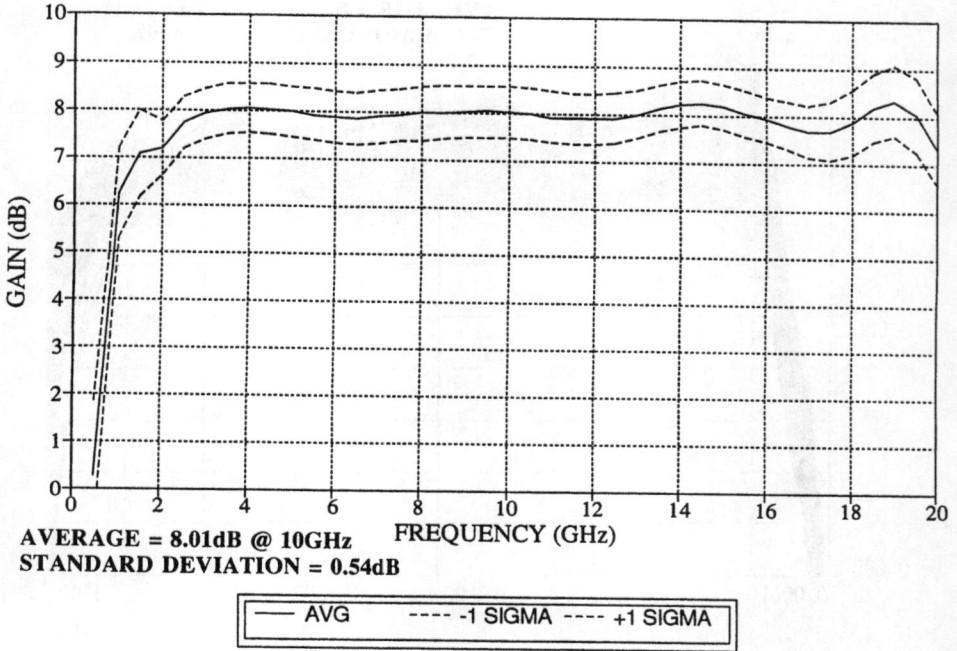

AVERAGE = 8.01dB @ 10GHz
STANDARD DEVIATION = 0.54dB

Figure 7.23 Measured small-signal gain variation.

7.7 CASE STUDY—A 7- TO 11-GHz LOW-NOISE MMIC AMPLIFIER

7.7.1 Comments

Most present-day MMIC-design practices overlook two important aspects required for accurate performance predictions: (1) adequate modeling of parasitic-coupling effects associated with circuit layout; and (2) statistical variations due to fabrication imprecision. Both of these design issues are addressed in this case study developed by Chris Henning at AT&T and presented in [15].

7.7.2 Preliminary Information

When a lumped design is distributed in a standard linear-circuit simulator, a number of the parasitics are taken into account. Junctions, steps, and bends are integrated into the simulation to account for electromagnetic parasitics found in these discontinuities, which are described by analytical expressions. Many of these discontinuity models are available elements in commercial CAD tools. Thus, the skill of

Percentage of 5000 samples (derived from 5000 trials).

←── 2 samples below   4 samples above ──→

AVERAGE = 8.08dB @ 10GHz
STANDARD DEVIATION = .44dB

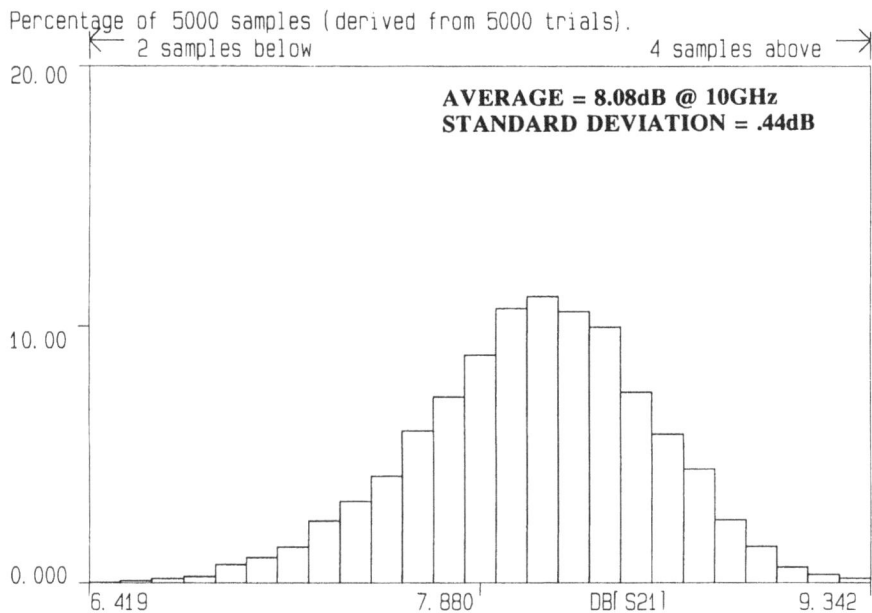

Figure 7.24 Performance histogram for gain.

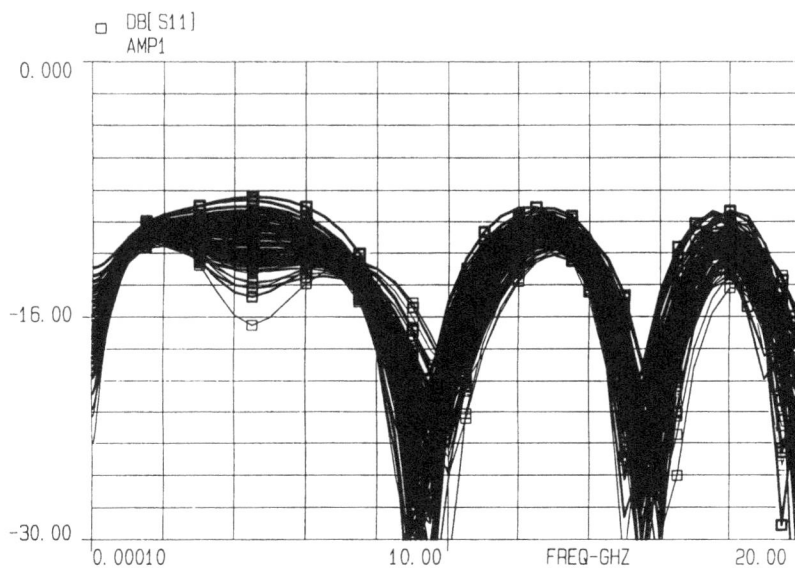

Figure 7.25 Predicted return loss variation.

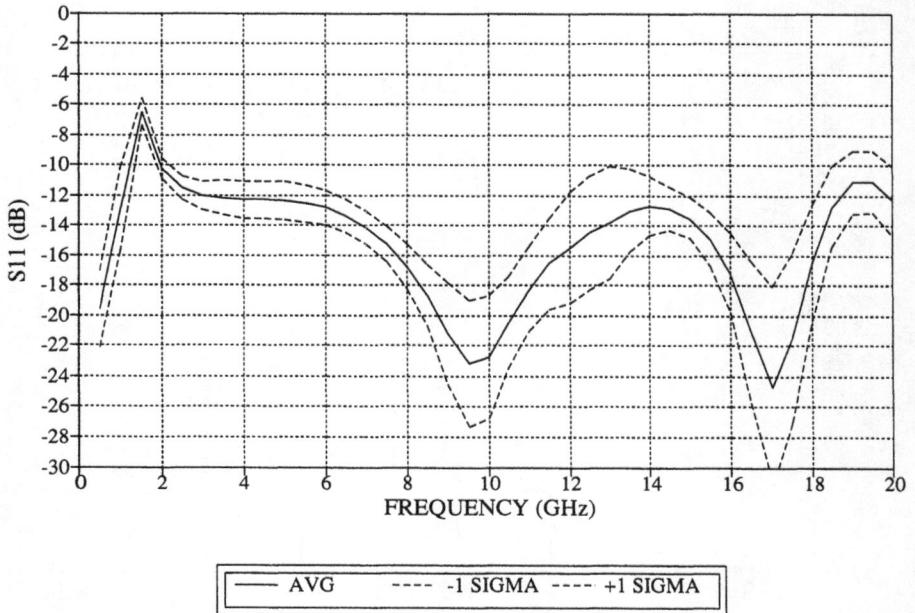

Figure 7.26 Measured input return loss variation.

a MMIC designer is often displayed by the ability to include the significant parasitics in the matching networks and neglect the insignificant parasitics.

There are, however, many more parasitic-coupling interactions than discontinuity parasitics in compacted MMIC circuits. By performing an electromagnetic simulation of the full circuit, a more accurate description of the parasitic interactions is included in the design process.

7.7.3 CAD-System Overview

The proposed CAD system for statistical design is shown diagrammatically in Figure 7.27. At the heart of the system is the ability to accurately model not only the statistical characteristics of the devices in the design, but also the effects of coupling between circuit elements (i.e., items 2 and 3 of Figure 7.27). For the implementation of item 2 in our system, we draw from previous works [10, 13, 15, 16] where it was illustrated that the so-called truth model is a good candidate for accurate and efficient statistical device modeling. In short, the truth model is an empirical-database model, and in the current application, it is extended to include device noise parameters. Implementation of item 3 of the CAD system utilizes

EMSim™ [2]. Electromagnetic simulation of full MMIC circuits using EMSim has been demonstrated and documented [17]. The following section presents a case study involving a three-stage LNA designed using the procedures outlined in Figure 7.27.

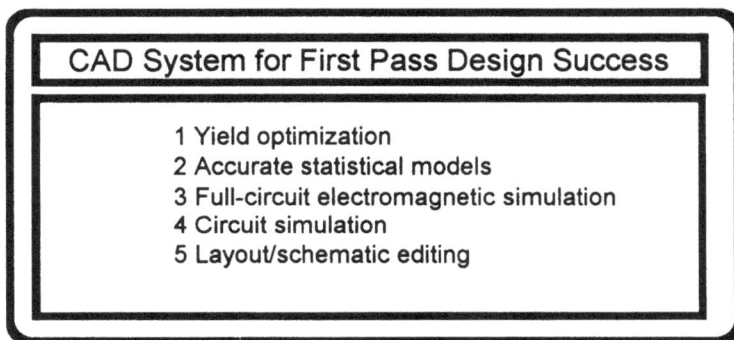

```
┌─────────────────────────────────────────────────────┐
│  ┌───────────────────────────────────────────────┐  │
│  │      CAD System for First Pass Design Success   │  │
│  └───────────────────────────────────────────────┘  │
│  ┌───────────────────────────────────────────────┐  │
│  │                                                 │  │
│  │        1 Yield optimization                     │  │
│  │        2 Accurate statistical models            │  │
│  │        3 Full-circuit electromagnetic simulation│  │
│  │        4 Circuit simulation                     │  │
│  │        5 Layout/schematic editing               │  │
│  │                                                 │  │
│  └───────────────────────────────────────────────┘  │
└─────────────────────────────────────────────────────┘
```

Figure 7.27 CAD system for first-pass design success.

7.7.4 A Three-Stage 7- to 11-GHz Low-Noise Amplifier

Circuit Description

A three-stage self-biased amplifier is designed to achieve the desired specifications for gain, noise figure, and input/output match over the 7- to 11-GHz frequency range, as shown in Figure 7.28. The first and second stages employ series feedback, while the final stage employs shunt feedback. The input match is designed to provide the optimum noise impedance, Γ_{opt}. The second stage, in conjunction with feedback and the interstage matching networks, ensures maximally flat gain and noise figure responses while maintaining low input/output VSWR. Active devices are represented by measured data, and the total chip dimensions are 1.48 by 1.42 by 0.1 mm.

Nominal Design

Once initial element values for the matching networks are determined, single-point optimization is performed using a minimax optimizer to flatten the responses and minimize the noise figure of the amplifier. Afterward, the design meets or exceeds the performance goals for a typical FET, but the average performance is lower, indicating a yield of only 25% (given the specifications in Figure 7.28). To achieve

LNA Specification:

Frequency range 7 - 11 GHz,
S21 = 19±1dB,
NF < 2.5 dB,
S11, S22 < -10 dB.

RF in

RF out

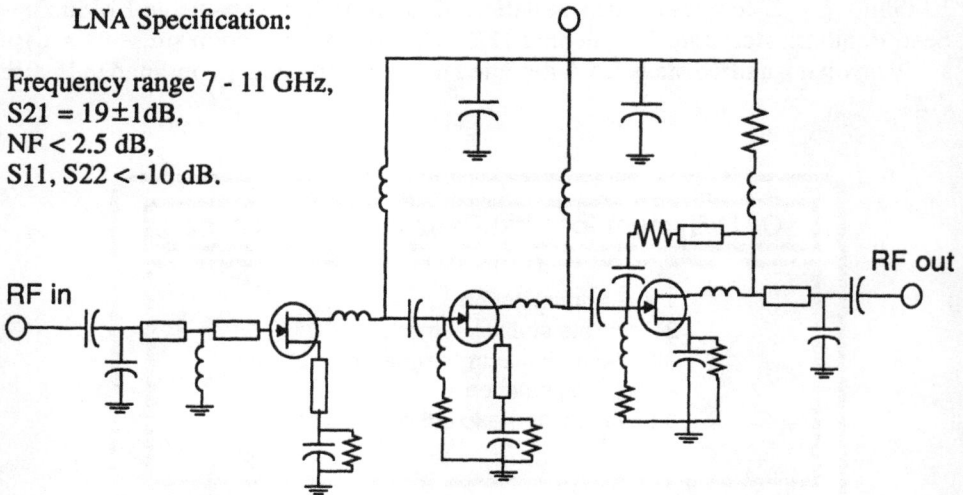

Figure 7.28 Schematic diagram of the 7- to 11-GHz LNA GaAs MMIC.

a more manufacturable design, we employ the statistical design and accurate statistical-modeling techniques resident within our proposed CAD system.

Robust Design

The methodology for robust design has two main parts: (1) Monte Carlo-based design-centering algorithms; and (2) accurate and efficient device statistical models. In this case study, the design-centering algorithms employed in Libra [2] are used.

As mentioned previously, a database model is the choice for the device statistical model. To construct this database model, both Scattering and noise parameters were measured on-wafer for 45 HFETs across five wafers from four lots. Based on empirical evidence, we believe this measurement database sufficiently represents process variation.

After designating designable variables, the automatic centering algorithm is used in conjunction with the linear simulator and the yield is optimized to better than 75% (according to the specifications of Figure 7.28). Figure 7.29 shows the statistical response of the LNA (a) before and (b) after design centering.

Statistical Electromagnetic Circuit Simulation

The final step in the design process involves electromagnetic simulation of the design-centered circuit. Figure 7.30 shows the layout of the LNA with the current

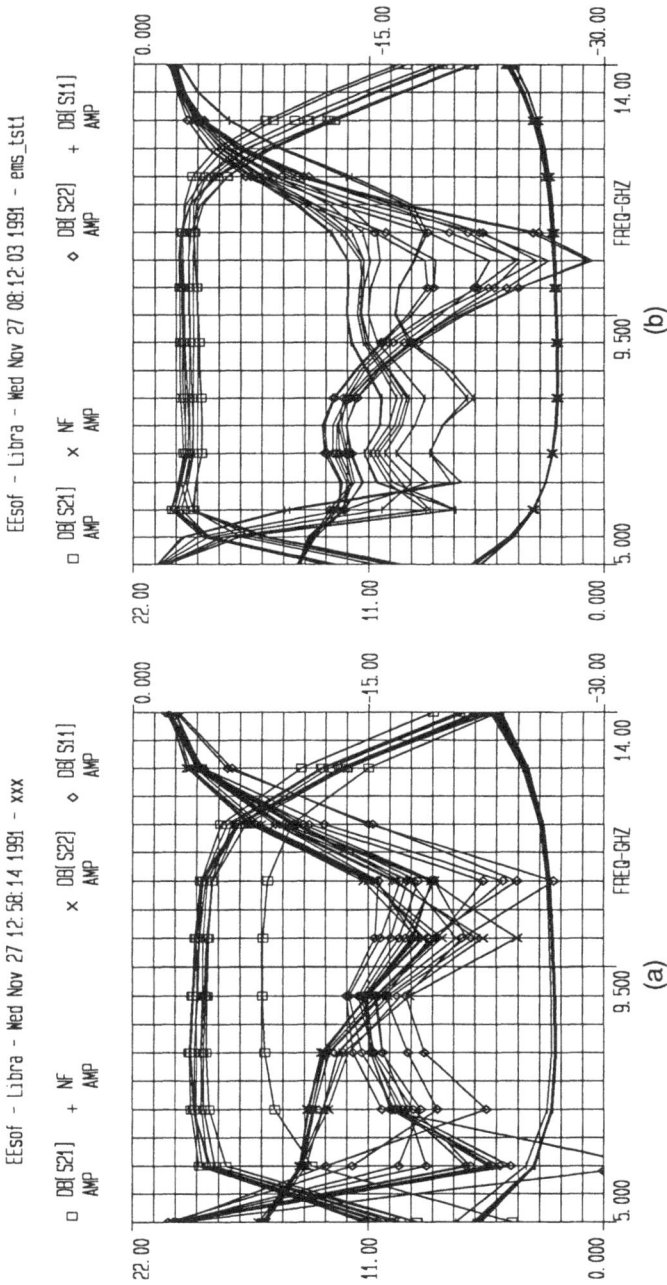

Figure 7.29 LNA statistical response (a) before and (b) after design centering.

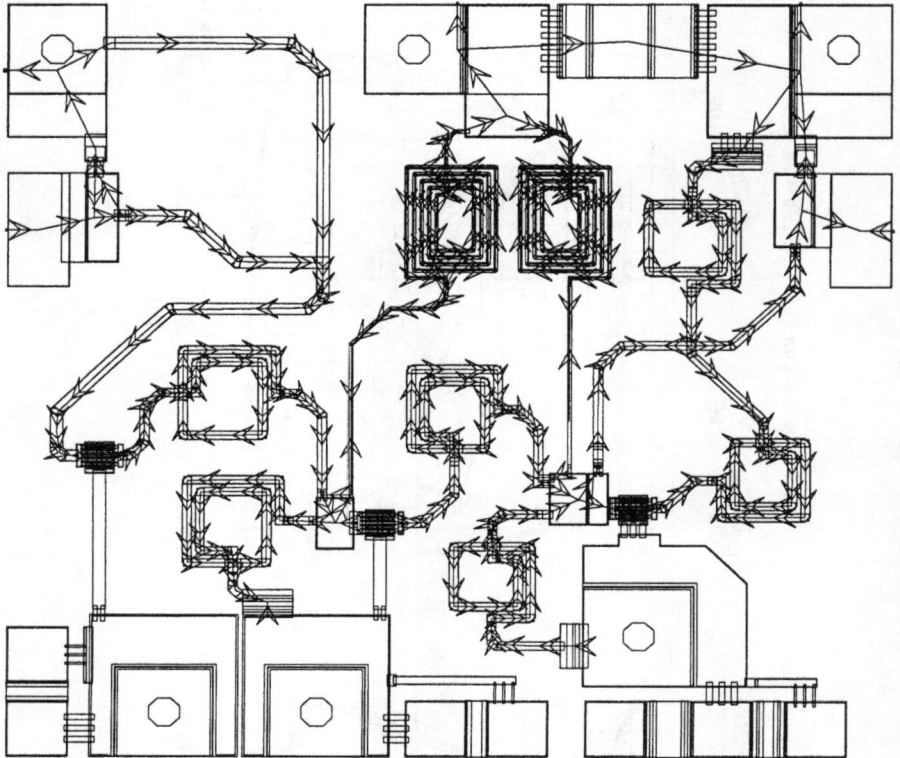

Figure 7.30 LNA patched for electromagnetic simulation.

segment patching scheme used in the moment-method-based electromagnetic simulation. Figure 7.31 shows the expected statistical response of the LNA where a 10-port EMSim S-parameter file is used in conjunction with the combined scattering and noise parameter truth model. (The simulation has 332 unknowns and takes about 37 minutes per frequency on a SPARCstation IPC with 12 megabytes of RAM.) Comparing Figure 7.31 to Figure 7.29(b), note that the high end of the frequency band has greater deviations, primarily due to the greater impact of parasitic coupling of the elements in the matching networks. There is a change in S22 throughout the band, and further investigation is required to isolate the main factors contributing to this change. Here again though, the difference is most likely due to parasitic coupling. The performance results given by the statistical electromagnetic simulation were deemed acceptable and the design is currently being fabricated. However, at press time, statistically meaningful data was not yet available, and these results will be reported elsewhere. Figure 7.32 shows a photograph of the MMIC LNA.

EEsof - Libra - Wed Nov 27 08:12:03 1991 - ems_tst1

□ DB[S21]   × NF   ◇ DB[S22]   + DB[S11]
  EMS   EMS   EMS   EMS

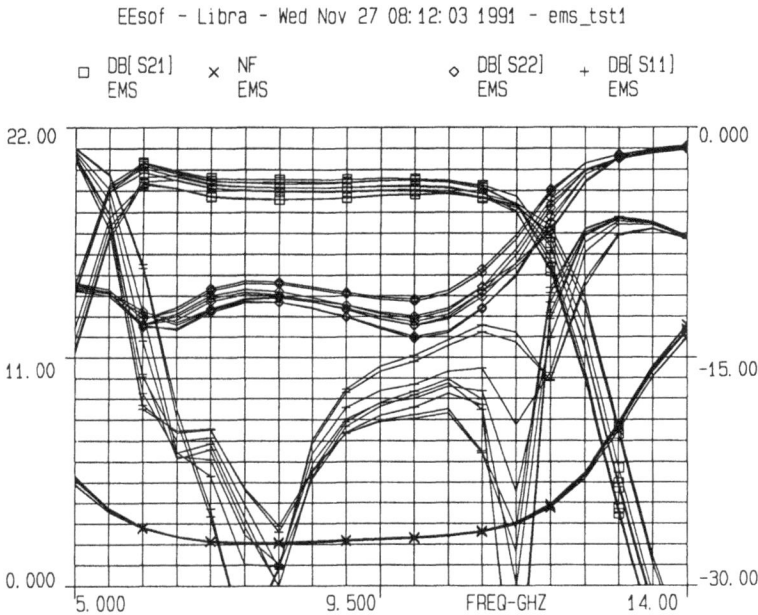

Figure 7.31 LNA statistical response from EMSim.

7.7.5 Summation

A demonstration of an advanced CAD system for statistical-design has been made. By combining statistical-design and modeling techniques with circuit-level electro-magnetic simulation, the engineer can submit designs to production with a high level of confidence.

7.8 CASE STUDY—DESIGN TO COST

7.8.1 Comments

This example of a statistical database approach to yield analysis was supplied by Joe Glaser and John Shioli of TRW.

7.8.2 Preliminary Information

Design to cost (DTC) provides a framework for analyzing the simulated effects of process variations upon circuit performance. What distinguishes DTC from other tools which attempt to model and predict circuit performance is DTC actually uses

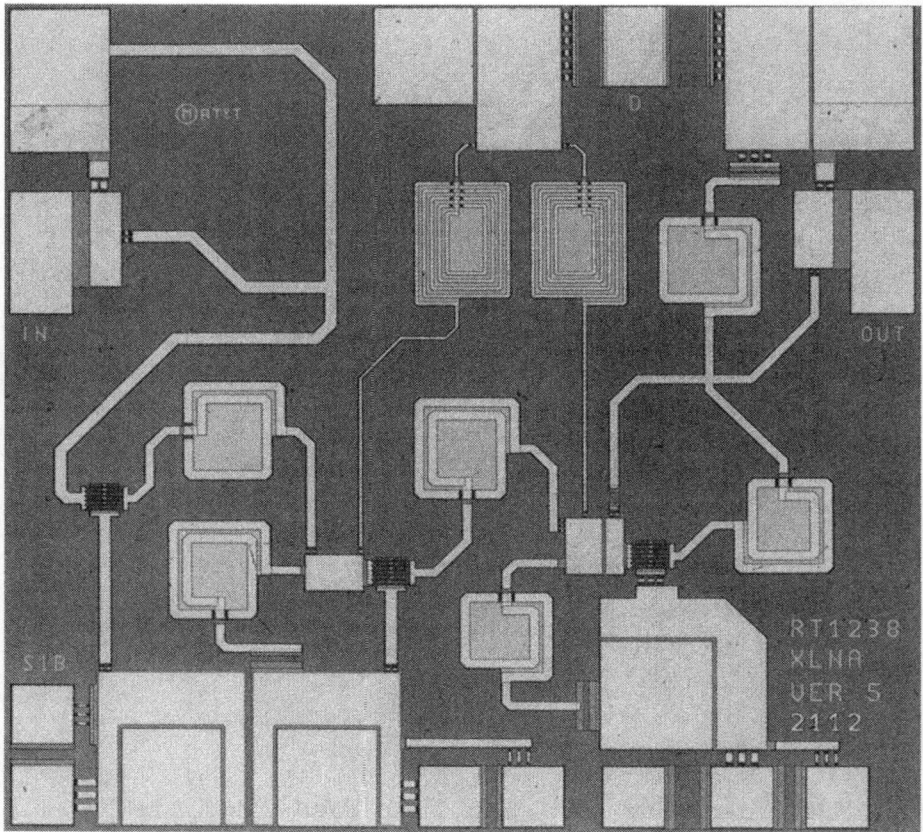

Figure 7.32 Photograph of the MMIC LNA.

fabrication and test data which represent the effects of process variation upon the circuit performance.

The main principle behind DTC is the concept of incorporating process-variation data into simulation of the circuit designs as a means of predicting how the process variations will influence the nominal values in the design, which will in turn effect the overall behavior of a circuit. The method for doing this is to use process-variation data in the form of discrete measurements (resistors, capacitors), parametric and S-parameter data (FETs, diodes, bipolar transistors). The parametric data is converted into standard large- and small-signal models that are stored with the passive data in the database. The passive data and active models can be retrieved from the database and substituted into the circuit netlist where each netlist represents a trial in the Monte Carlo experiment. The range of variation is modeled by taking samples from the selected data population. This approach was chosen due to the nonlinear dependent nature of the parasitic elements within the active devices which prevented more simplistic statistical models from being useful.

7.8.3 Design-to-Cost Framework

DTC automates the simulation process by querying the designer to provide a specification of the circuit to be simulated, the range of fabrication data and models upon which the simulations will be based, and the number of simulations. The designer can interactively select which of the given lot of wafers will form the basis of the yield model and the number of simulations that will influence the statistical accuracy of the predicted yields produced by DTC. The designer can vary the range of lots to predict the best and worst case interpretation of the process as a means of predicting how this variation will effect the circuit behavior. DTC scans the circuit netlist for passive and active elements, which are substituted by the appropriate passive or active models represented by the range of data chosen by the designer. The simulation results are collated and displayed to show the effect of the process variation upon the circuit-performance behavior. In this way, the simulation results can be compared directly to the designer's specifications, which are used to determine the RF yield and chip die cost.

To effectively design for production, the designer must optimize the circuit performance so that nominal circuit performance is obtained from the nominal process variations. DTC enables circuit designers to forecast the yield and cost of their designs in a production environment and to trade performance, cost and sensitivity to obtain an optimal design. This capability is critical to reducing the cost of the development process and to improving the design accuracy.

Selection of circuits for analysis can be accomplished by choosing local netlist files or querying the database for circuit types (amplifier, mixer etc.). After the circuit type has been chosen, a list of circuits is supplied to the designer for selection. This is shown in Figure 7.33.

Once the circuit has been selected, the designer can chose a range of passive data and device models as a function of the technology (MESFET, HEMT, HBT, etc.), production date, lot, or wafer that will represent the sample of models for the simulation. This is shown in Figure 7.34.

The designer can vary the confidence level by choosing from one-, two-, or three-sigma distributions. By selecting one-sigma the designer will have only a 68% confidence that the theoretical yield will actually be met. The number of simulations will effect the calculation of the error estimate. For a given confidence level, more simulations will be needed to decrease the error estimate to obtain a higher degree of accuracy. This is shown in Figure 7.35.

7.8.4 Results

The results of the analysis show the RF specification entered by the designer and the cost-to-yield predictions. The results of the simulations are available to the

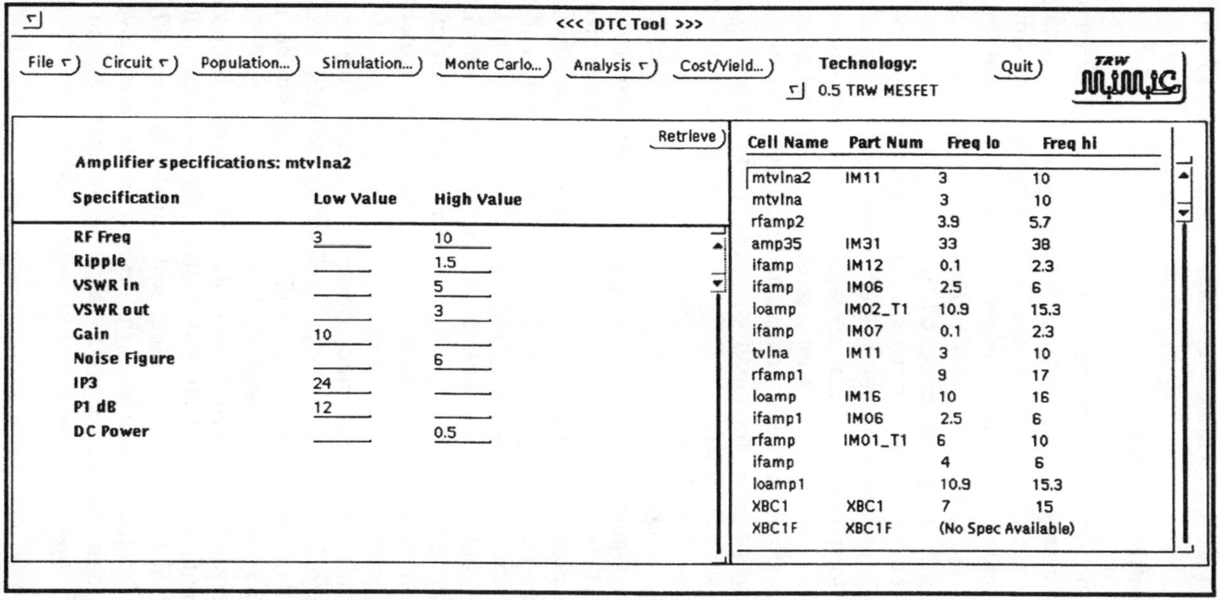

256

File ▾ Circuit ▾ Population... Simulation... Monte Carlo... Analysis ▾ Cost/Yield... Technology: Quit *TRW*

<<< DTC Tool >>>

Technology: 0.5 TRW MESFET

Amplifier specifications: mtvlna2

Specification	Low Value	High Value
RF Freq	3	10
Ripple		1.5
VSWR in		5
VSWR out		3
Gain	10	
Noise Figure		6
IP3	24	
P1 dB	12	
DC Power		0.5

Retrieve

Cell Name	Part Num	Freq lo	Freq hi
mtvlna2	IM11	3	10
mtvlna		3	10
rfamp2		3.9	5.7
amp35	IM31	33	38
lfamp	IM12	0.1	2.3
ifamp	IM06	2.5	6
loamp	IMO2_T1	10.9	15.3
ifamp	IM07	0.1	2.3
tvlna	IM11	3	10
rfamp1		9	17
loamp	IM16	10	16
ifamp1	IM06	2.5	6
rfamp	IMO1_T1	6	10
ifamp		4	6
loamp1		10.9	15.3
XBC1	XBC1	7	15
XBC1F	XBC1F	(No Spec Available)	

Figure 7.33 Selection of circuits for analysis.

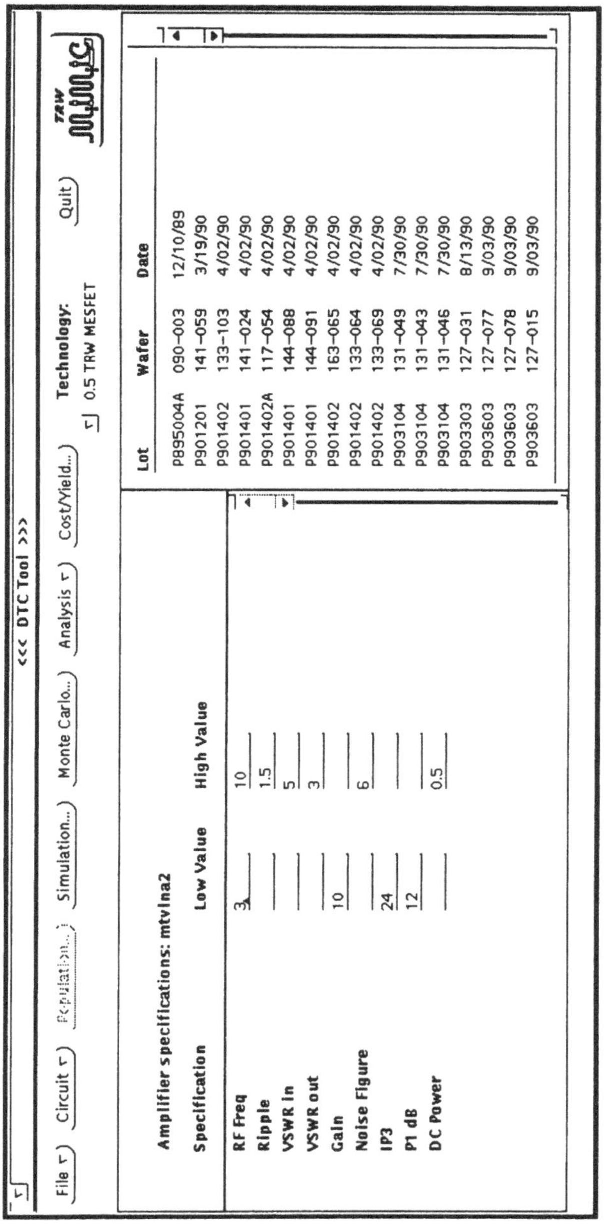

<<< DTC Tool >>>

File ▼ | Circuit ▼ | Population... | Simulation... | Monte Carlo... | Analysis ▼ | Cost/Yield... | Technology: | Quit

Technology: 0.5 TRW MESFET

Amplifier specifications: mtvlna2

Specification	Low Value	High Value
RF Freq	3	10
Ripple		1.5
VSWR In		5
VSWR out		3
Gain	10	
Noise Figure		6
IP3	24	
P1 dB	12	
DC Power		0.5

Lot	Wafer	Date
PB95004A	090–003	12/10/89
P901201	141–059	3/19/90
P901402	133–103	4/02/90
P901401	141–024	4/02/90
P901402A	117–054	4/02/90
P901401	144–088	4/02/90
P901401	144–091	4/02/90
P901402	163–065	4/02/90
P901402	133–064	4/02/90
P901402	133–069	4/02/90
P903104	131–049	7/30/90
P903104	131–043	7/30/90
P903104	131–046	7/30/90
P903303	127–031	8/13/90
P903603	127–077	9/03/90
P903603	127–078	9/03/90
P903603	127–015	9/03/90

Figure 7.34 Selection of passive data and device models.

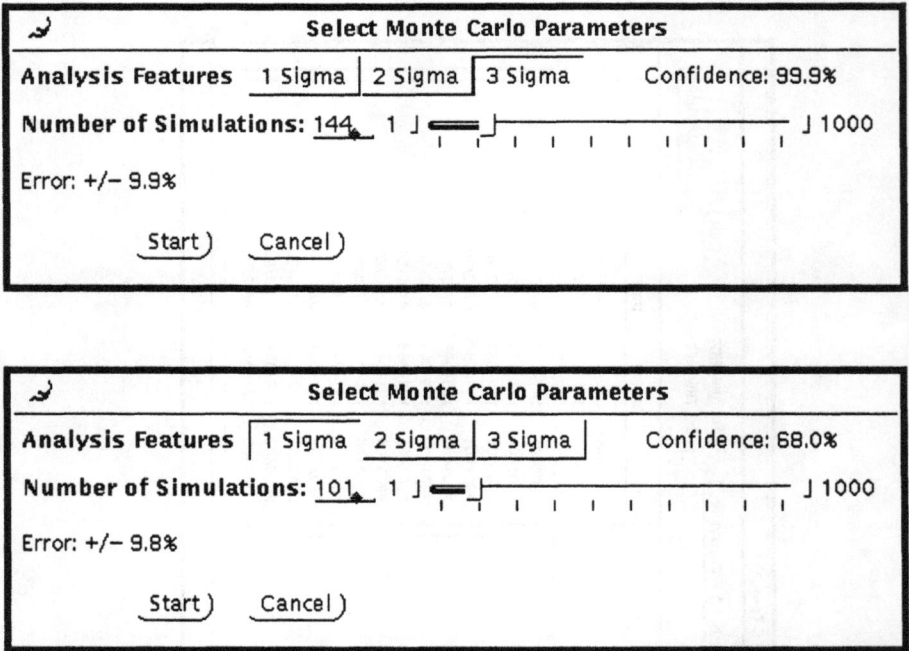

Figure 7.35 Selection of the yield-estimate confidence-level.

designer in graphic form for a direct comparison of yield-to-circuit performance. This is shown in Figure 7.36.

REFERENCES

[1] Keith Cobler, "Statistical Design Improves Reliability and Manufacturing Yields of RF Circuits," RF Design, April 1992, pp. 24-35.

[2] EEsof Inc., Westlake Village, CA.

[3] G.D. Vendelin, A.M. Pavio and U.L. Rohde, Microwave Circuit Design Using Linear and Non-linear Techniques, J.W Wiley & Sons, New York, NY, 1990, pp. 283-288.

[4] J.W. Bandler, R.M. Biernacki, et al., "Gradient Quadratic Approximation Scheme for Yield-driven Design," Proceedings of the IEEE MTT-S International Microwave Symposium, Boston, MA, May 1991, pp. 1197-1200.

[5] J. Purviance and M. Meehan, "CAD for Statistical Analysis and Design of Microwave Circuits," Int. Journal of Microw. and Mill.-Wave Comp. Aided Des., Vol. 1, No. 1, January 1991.

[6] U.L. Rohde, M. Anton, D. Gabbay, R. Gilmore, "MMIC Workstations for the 1990's," Microwave Journal, State of the Art Reference, pp. 51-77, sup. Sept 1989.

[7] J. Baprawski, N. Kamaglekar, "Omnisys: Simulator for the Microwave System Designer", Microwave Journal, International Edition, pp. 379-387, May 1988.

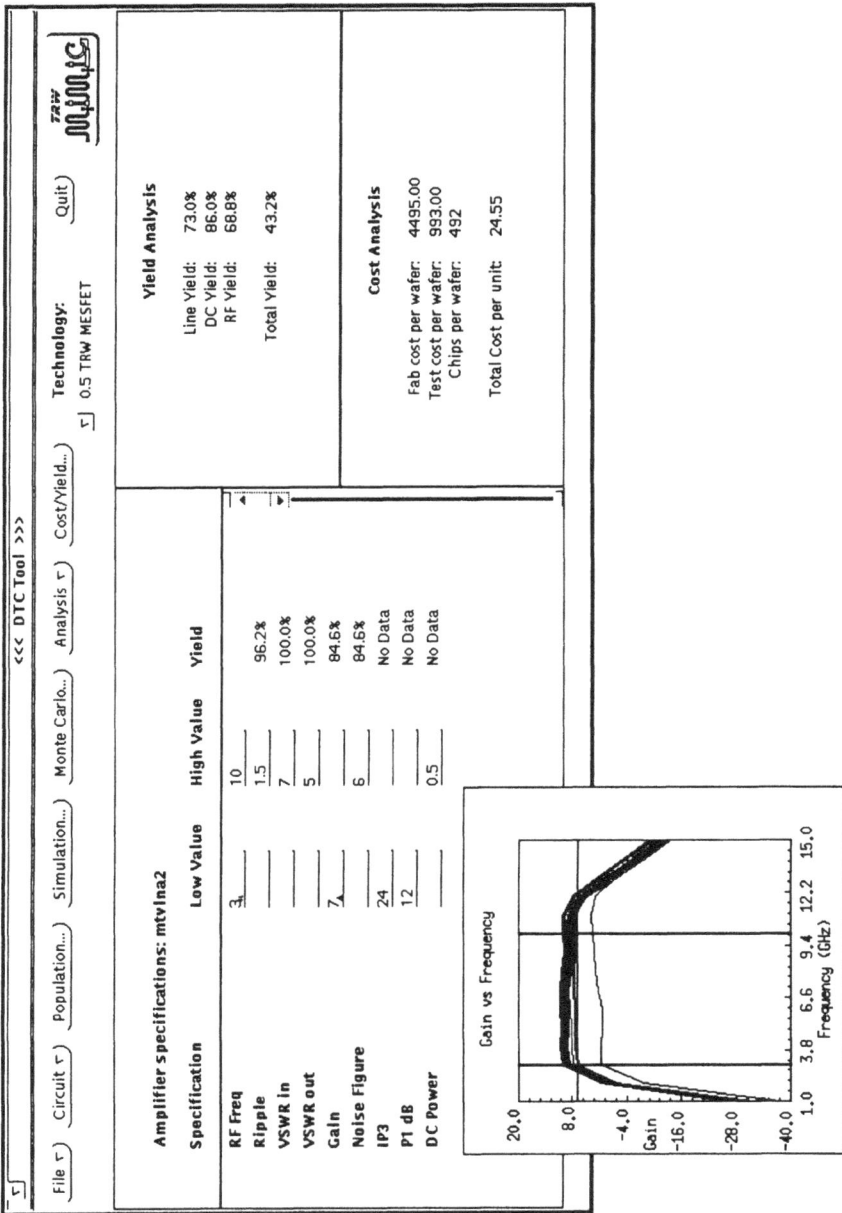

Figure 7.36 The results of the simulations.

[8] R. Cooke and J. Purviance, "Statistical Design for Microwave Systems," Proceedings of the IEEE MTT-S International Microwave Symposium, Boston, MA, May 1991, pp. 679–682.

[9] OMNISYS Applications Manual, version 1.1, October 1989, pp. converter 1,2,3.

[10] M. Meehan, T. Wandinger, and D.A. Fisher, "Accurate Design Centering and Yield Prediction using the "Truth Model"," IEEE MTT-S International Microwave Symposium, Boston, MA., June 1991, pp. 1201-1204.

[11] Watkins-Johnson Company, Palo Alto, CA.

[12] D.L. Allen, J. Beall, and M. King, "Small-Signal RF Yield Analysis Of MMIC Circuits Based On Physical Device Parameters" 1992 IEEE MTT-S Digest, pp. 1473–1476.

[13] Purviance, J., M. Meehan, and D. Collins, "Properties of FET parameter statistical data bases," IEEE Int. Microwave Symposium Digest (Dallas,TX), 1990, pp.567-570.

[14] Campbell, L., J. Purviance, and C. Potratz, "Statistical Interpolation of FET Data Base Measurements," 1991 IEEE MTT-S Digest, pp. 201-204.

[15] M. Meehan, L. Campbell, "Statistical Techniques for Objective Characterization of Microwave Device Statistical Data," submitted for publication to the IEEE MTT-S International Microwave Symposium, Boston, MA, June 1991.

[16] M.D. Meehan, D.M. Collins, "Investigation of the GaAs FET Model to Assess its Applicability to Design Centering and Yield Estimation," *EEsof International Development Report*, December 1987.

[17] P. Draxler, G.E. Howard, Y.L. Chow, "Mixed Spectral/Spatial Domain Moment Method Simulation of Components and Circuits," *Proc. 21st European Microwave Conference*, Stuttgart, Germany, September 1991, pp. 1284–1289.

Appendix A
Monte Carlo Confidence Interval Tables

MONTE CARLO

Sample Size, Accuracy and Confidence Level

Confidence In Yield Estimate Y_{est}:

$$|Y_{est} - Y| \le Z_c \sqrt{\frac{Y(1 - Y)}{N}}$$

Y = actual yield $(0 \le Y \le 1)$
Y_{est} = Monte Carlo Yield Estimate $(0 \le Y_{est} \le 1)$

Note: the yields here are fractions and not percentages

N = Number of Monte Carlo Samples
Z_c = Confidence Level Variable

assumes N is large enough that:

$$Y > .5 \text{ and } N(1 - Y) > 5$$
or
$$Y < .5 \text{ and } NY > 5$$

Confidence Level	99%	95%	80%	50%
Z_c	2.575	1.96	1.28	.6745

MONTE CARLO

Sample Size, Accuracy and Confidence Level

$$|Y_{est} - Y| \le Z_c\sqrt{\frac{Y(1 - Y)}{N}}$$

Confidence Level	Z_c	Y	N	solve for $|Y_{est} - Y|$
99%	2.85	.80	1000	.036
80%	1.28	.6	500	.028
95%	1.96	.7	2000	.020
80%	1.28	.7	2000	.013
95%	1.96	.7	200	.20

In practice, Y_{est}, the estimated yield, is substituted for Y, the actual yield, in the above table.

For most design confidence levels of 95 to 99% are adequate. The uncertainty in component statistics and circuit models can make higher confidence levels meaningless.

Confidence = 68.3%   Actual Yield = 90%

Error ± %	Estimated % Yield Low	High	Number of Trials
1.0	89.00	91.00	900
2.0	88.00	92.00	225
3.0	87.00	93.00	100
4.0	86.00	94.00	56
5.0	85.00	95.00	36
6.0	84.00	96.00	25
7.0	83.00	97.00	18
8.0	82.00	98.00	14
9.0	81.00	99.00	11
10.0	80.00	100.00	9

Confidence = 95%   Actual Yield = 90%

Error ± %	Estimated % Yield Low	High	Number of Trials
1.0	89.00	91.00	3457
2.0	88.00	92.00	864
3.0	87.00	93.00	384
4.0	86.00	94.00	216
5.0	85.00	95.00	138
6.0	84.00	96.00	96
7.0	83.00	97.00	70
8.0	82.00	98.00	54
9.0	81.00	99.00	42
10.0	80.00	100.00	34

Confidence = 99%   Actual Yield = 90%

Error ± %	Estimated % Yield Low	High	Number of Trials
1.0	89.00	91.00	5967
2.0	88.00	92.00	1491
3.0	87.00	93.00	663
4.0	86.00	94.00	372
5.0	85.00	95.00	238
6.0	84.00	96.00	165
7.0	83.00	97.00	121
8.0	82.00	98.00	93
9.0	81.00	99.00	73
10.0	80.00	100.00	59

Confidence = 68.3%   Actual Yield = 80%

| Error | Estimated % Yield | | Number |
± %	Low	High	of Trials
1.0	79.00	81.00	1600
2.0	78.00	82.00	400
3.0	77.00	83.00	177
4.0	76.00	84.00	100
5.0	75.00	85.00	64
6.0	74.00	86.00	44
7.0	73.00	87.00	32
8.0	72.00	88.00	25
9.0	71.00	89.00	19
10.0	70.00	90.00	16

Confidence = 95%   Actual Yield = 80%

| Error | Estimated % Yield | | Number |
± %	Low	High	of Trials
1.0	79.00	81.00	6146
2.0	78.00	82.00	1536
3.0	77.00	83.00	682
4.0	76.00	84.00	384
5.0	75.00	85.00	245
6.0	74.00	86.00	170
7.0	73.00	87.00	125
8.0	72.00	88.00	96
9.0	71.00	89.00	75
10.0	70.00	90.00	61

Confidence = 99%   Actual Yield = 80%

| Error | Estimated % Yield | | Number |
± %	Low	High	of Trials
1.0	79.00	81.00	10609
2.0	78.00	82.00	2652
3.0	77.00	83.00	1178
4.0	76.00	84.00	663
5.0	75.00	85.00	424
6.0	74.00	86.00	294
7.0	73.00	87.00	216
8.0	72.00	88.00	165
9.0	71.00	89.00	130
10.0	70.00	90.00	106

Confidence = 68.3%   Actual Yield = 70%

| Error | Estimated % Yield | | Number |
± %	Low	High	of Trials
1.0	69.00	71.00	2100
2.0	68.00	72.00	525
3.0	67.00	73.00	233
4.0	66.00	74.00	131
5.0	65.00	75.00	84
6.0	64.00	76.00	58
7.0	63.00	77.00	42
8.0	62.00	78.00	32
9.0	61.00	79.00	25
10.0	60.00	80.00	21

Confidence = 95%   Actual Yield = 70%

| Error | Estimated % Yield | | Number |
± %	Low	High	of Trials
1.0	69.00	71.00	8067
2.0	68.00	72.00	2016
3.0	67.00	73.00	896
4.0	66.00	74.00	504
5.0	65.00	75.00	322
6.0	64.00	76.00	224
7.0	63.00	77.00	164
8.0	62.00	78.00	126
9.0	61.00	79.00	99
10.0	60.00	80.00	80

Confidence = 99%   Actual Yield = 70%

| Error | Estimated % Yield | | Number |
± %	Low	High	of Trials
1.0	69.00	71.00	13924
2.0	68.00	72.00	3481
3.0	67.00	73.00	1547
4.0	66.00	74.00	870
5.0	65.00	75.00	556
6.0	64.00	76.00	386
7.0	63.00	77.00	284
8.0	62.00	78.00	217
9.0	61.00	79.00	171
10.0	60.00	80.00	139

Confidence = 68.3%   Actual Yield = 60%

| Error | Estimated % Yield | | Number |
± %	Low	High	of Trials
1.0	59.00	61.00	2400
2.0	58.00	62.00	600
3.0	57.00	63.00	266
4.0	56.00	64.00	150
5.0	55.00	65.00	96
6.0	54.00	66.00	66
7.0	53.00	67.00	48
8.0	52.00	68.00	37
9.0	51.00	69.00	29
10.0	50.00	70.00	24

Confidence = 95%   Actual Yield = 60%

| Error | Estimated % Yield | | Number |
± %	Low	High	of Trials
1.0	59.00	61.00	9219
2.0	58.00	62.00	2304
3.0	57.00	63.00	1024
4.0	56.00	64.00	576
5.0	55.00	65.00	368
6.0	54.00	66.00	256
7.0	53.00	67.00	188
8.0	52.00	68.00	144
9.0	51.00	69.00	113
10.0	50.00	70.00	92

Confidence = 99%   Actual Yield = 60%

| Error | Estimated % Yield | | Number |
± %	Low	High	of Trials
1.0	59.00	61.00	15913
2.0	58.00	62.00	3978
3.0	57.00	63.00	1768
4.0	56.00	64.00	994
5.0	55.00	65.00	636
6.0	54.00	66.00	442
7.0	53.00	67.00	324
8.0	52.00	68.00	248
9.0	51.00	69.00	196
10.0	50.00	70.00	159

Confidence = 68.3%   Actual Yield = 50%

Error ± %	Estimated % Yield		Number of Trials
	Low	*High*	
1.0	49.00	51.00	2500
2.0	48.00	52.00	625
3.0	47.00	53.00	277
4.0	46.00	54.00	156
5.0	45.00	55.00	100
6.0	44.00	56.00	69
7.0	43.00	57.00	51
8.0	42.00	58.00	39
9.0	41.00	59.00	30
10.0	40.00	60.00	25

Confidence = 95%   Actual Yield = 50%

Error ± %	Estimated % Yield		Number of Trials
	Low	*High*	
1.0	49.00	51.00	9604
2.0	48.00	52.00	2401
3.0	47.00	53.00	1067
4.0	46.00	54.00	600
5.0	45.00	55.00	384
6.0	44.00	56.00	266
7.0	43.00	57.00	196
8.0	42.00	58.00	150
9.0	41.00	59.00	118
10.0	40.00	60.00	96

Confidence = 99%   Actual Yield = 50%

Error ± %	Estimated % Yield		Number of Trials
	Low	*High*	
1.0	49.00	51.00	16576
2.0	48.00	52.00	4144
3.0	47.00	53.00	1841
4.0	46.00	54.00	1036
5.0	45.00	55.00	663
6.0	44.00	56.00	460
7.0	43.00	57.00	338
8.0	42.00	58.00	259
9.0	41.00	59.00	204
10.0	40.00	60.00	165

Confidence = 68.3%   Actual Yield = 40%

| Error | Estimated % Yield | | Number |
± %	Low	High	of Trials
1.0	39.00	41.00	2400
2.0	38.00	42.00	600
3.0	37.00	43.00	266
4.0	36.00	44.00	150
5.0	35.00	45.00	96
6.0	34.00	46.00	66
7.0	33.00	47.00	48
8.0	32.00	48.00	37
9.0	31.00	49.00	29
10.0	30.00	50.00	24

Confidence = 95%   Actual Yield = 40%

| Error | Estimated % Yield | | Number |
± %	Low	High	of Trials
1.0	39.00	41.00	9219
2.0	38.00	42.00	2304
3.0	37.00	43.00	1024
4.0	36.00	44.00	576
5.0	35.00	45.00	368
6.0	34.00	46.00	256
7.0	33.00	47.00	188
8.0	32.00	48.00	144
9.0	31.00	49.00	113
10.0	30.00	50.00	92

Confidence = 99%   Actual Yield = 40%

| Error | Estimated % Yield | | Number |
± %	Low	High	of Trials
1.0	39.00	41.00	15913
2.0	38.00	42.00	3978
3.0	37.00	43.00	1768
4.0	36.00	44.00	994
5.0	35.00	45.00	636
6.0	34.00	46.00	442
7.0	33.00	47.00	324
8.0	32.00	48.00	248
9.0	31.00	49.00	196
10.0	30.00	50.00	159

Confidence = 68.3%   Actual Yield = 30%

| Error | Estimated % Yield | | Number |
± %	Low	High	of Trials
1.0	29.00	31.00	2100
2.0	28.00	32.00	525
3.0	27.00	33.00	233
4.0	26.00	34.00	131
5.0	25.00	35.00	84
6.0	24.00	36.00	58
7.0	23.00	37.00	42
8.0	22.00	38.00	32
9.0	21.00	39.00	25
10.0	20.00	40.00	21

Confidence = 95%   Actual Yield = 30%

| Error | Estimated % Yield | | Number |
± %	Low	High	of Trials
1.0	29.00	31.00	8067
2.0	28.00	32.00	2016
3.0	27.00	33.00	896
4.0	26.00	34.00	504
5.0	25.00	35.00	322
6.0	24.00	36.00	224
7.0	23.00	37.00	164
8.0	22.00	38.00	126
9.0	21.00	39.00	99
10.0	20.00	40.00	80

Confidence = 99%   Actual Yield = 30%

| Error | Estimated % Yield | | Number |
± %	Low	High	of Trials
1.0	29.00	31.00	13924
2.0	28.00	32.00	3481
3.0	27.00	33.00	1547
4.0	26.00	34.00	870
5.0	25.00	35.00	556
6.0	24.00	36.00	386
7.0	23.00	37.00	284
8.0	22.00	38.00	217
9.0	21.00	39.00	171
10.0	20.00	40.00	139

Confidence = 68.3%   Actual Yield = 20%

| Error ± % | Estimated % Yield | | Number |
	Low	High	of Trials
1.0	19.00	21.00	1600
2.0	18.00	22.00	400
3.0	17.00	23.00	177
4.0	16.00	24.00	100
5.0	15.00	25.00	64
6.0	14.00	26.00	44
7.0	13.00	27.00	32
8.0	12.00	28.00	25
9.0	11.00	29.00	19
10.0	10.00	30.00	16

Confidence = 95%   Actual Yield = 20%

| Error ± % | Estimated % Yield | | Number |
	Low	High	of Trials
1.0	19.00	21.00	6146
2.0	18.00	22.00	1536
3.0	17.00	23.00	682
4.0	16.00	24.00	384
5.0	15.00	25.00	245
6.0	14.00	26.00	170
7.0	13.00	27.00	125
8.0	12.00	28.00	96
9.0	11.00	29.00	75
10.0	10.00	30.00	61

Confidence = 99%   Actual Yield = 20%

| Error ± % | Estimated % Yield | | Number |
	Low	High	of Trials
1.0	19.00	21.00	10609
2.0	18.00	22.00	2652
3.0	17.00	23.00	1178
4.0	16.00	24.00	663
5.0	15.00	25.00	424
6.0	14.00	26.00	294
7.0	13.00	27.00	216
8.0	12.00	28.00	165
9.0	11.00	29.00	130
10.0	10.00	30.00	106

Confidence = 68.3%   Actual Yield = 10%

| Error | Estimated % Yield | | Number |
± %	Low	High	of Trials
1.0	19.00	21.00	899
2.0	18.00	22.00	224
3.0	17.00	23.00	100
4.0	16.00	24.00	56
5.0	15.00	25.00	36
6.0	14.00	26.00	25
7.0	13.00	27.00	18
8.0	12.00	28.00	14
9.0	11.00	29.00	11
10.0	10.00	30.00	9

Confidence = 95%   Actual Yield = 10%

| Error | Estimated % Yield | | Number |
± %	Low	High	of Trials
1.0	19.00	21.00	3457
2.0	18.00	22.00	864
3.0	17.00	23.00	384
4.0	16.00	24.00	216
5.0	15.00	25.00	138
6.0	14.00	26.00	96
7.0	13.00	27.00	70
8.0	12.00	28.00	54
9.0	11.00	29.00	42
10.0	10.00	30.00	34

Confidence = 99%   Actual Yield = 10%

| Error | Estimated % Yield | | Number |
± %	Low	High	of Trials
1.0	19.00	21.00	5967
2.0	18.00	22.00	1491
3.0	17.00	23.00	663
4.0	16.00	24.00	372
5.0	15.00	25.00	238
6.0	14.00	26.00	165
7.0	13.00	27.00	121
8.0	12.00	28.00	93
9.0	11.00	29.00	73
10.0	10.00	30.00	59

Index

The Artech House Microwave Library

LOSLIN: Lossy Line Calculation Software and User's Manual, Fred E. Gardiol

Lossy Transmission Lines, Fred E. Gardiol

Low-Angle Microwave Propagation: Physics and Modeling, Adolf Giger

Low Phase Noise Microwave Oscillator Design, Robert G. Rogers

MATCHNET: Microwave Matching Networks Synthesis, Stephen V. Sussman-Fort

Materials Handbook for Hybrid Microelectronics, J.A. King, ed.

Matrix Parameters for Multiconductor Transmission Lines: Software and User's Manual, A.R. Djordjevic, et al.

MIC and MMIC Amplifier and Oscillator Circuit Design, Allen Sweet

Microelectronic Reliability, Volume I: Reliability, Test, and Diagnostics, Edward B. Hakim, ed.

Microelectronic Reliability, Volume II: Integrity Assessment and Assurance, Emiliano Pollino, ed.

Microstrip Lines and Slotlines, D.C. Gupta, R. Garg, and I.J. Bahl

Microwave and RF Circuits: Analysis, Synthesis, and Design, Max Medley

Microwave and RF Component and Subsystem Manufacturing Technology, Heriot-Watt University

Microwave Circulator Design, Douglas K. Linkhart

Microwave Engineers' Handbook: 2 Volume Set, Theodore Saad, ed.

Microwave Materials and Fabrication Techniques, Second Edition, Thomas S. Laverghetta

Microwave MESFETs and HEMTs, J. Michael Golio, et al.

Microwave and Millimeter Wave Heterostructure Transistors and Applicatons, F. Ali, ed.

Microwave and Millimeter Wave Phase Shifters, Volume I: Dielectric and Ferrite Phase Shifters, S. Koul and B. Bhat

Microwave and Millimeter Wave Phase Shifters, Volume II: Semiconductor and Delay Line Phase Shifters, S. Koul and B. Bhat

Microwave Mixers, Second Edition, Stephen Maas

Microwave Transmission Design Data, Theodore Moreno

Microwave Transition Design, Jamal S. Izadian and Shahin M. Izadian

Microwave Transmission Line Couplers, J.A.G. Malherbe

Microwave Tubes, A.S. Gilmour, Jr.

Microwaves: Industrial, Scientific, and Medical Applications, J. Thuery

Microwaves Made Simple: Principles and Applicatons, Stephen W. Cheung, Frederick H. Levien, et al.

MMIC Design: GaAs FETs and HEMTs, Peter H. Ladbrooke

Modern GaAs Processing Techniques, Ralph Williams

Modern Microwave Measurements and Techniques, Thomas S. Laverghetta

Monolithic Microwave Integrated Circuits: Technology and Design, Ravender
    Goyal, et al.

Nonlinear Microwave Circuits, Stephen A. Maas

Optical Control of Microwave Devices, Rainee N. Simons

PLL: Linear Phase-Locked Loop Control Systems Analysis Software and User's Manual,
    Eric L. Unruh

*Scattering Parameters of Microwave Networks with Multiconductor Transmission Lines:
    Software and User's Manual*, A.R. Djordjevic, et al.

Solid-State Microwave Power Oscillator Design, Eric Holzman and Ralston Robertson

Stripline Circuit Design, Harlan Howe, Jr.

Terrestrial Digital Microwave Communications, Ferdo Ivanek, et al.

*Time-Domain Response of Multiconductor Transmission Lines: Software and User's
    Manual*, A.R. Djordjevic, et al.

Transmission Line Design Handbook, Brian C. Waddell

www.ingramcontent.com/pod-product-compliance
Lightning Source LLC
Chambersburg PA
CBHW021429180326
41458CB00001B/192